Studies in Computational Intelligence

Volume 719

Series editor

Janusz Kacprzyk, Polish Academy of Sciences, Warsaw, Poland
e-mail: kacprzyk@ibspan.waw.pl

About this Series

The series "Studies in Computational Intelligence" (SCI) publishes new developments and advances in the various areas of computational intelligence—quickly and with a high quality. The intent is to cover the theory, applications, and design methods of computational intelligence, as embedded in the fields of engineering, computer science, physics and life sciences, as well as the methodologies behind them. The series contains monographs, lecture notes and edited volumes in computational intelligence spanning the areas of neural networks, connectionist systems, genetic algorithms, evolutionary computation, artificial intelligence, cellular automata, self-organizing systems, soft computing, fuzzy systems, and hybrid intelligent systems. Of particular value to both the contributors and the readership are the short publication timeframe and the worldwide distribution, which enable both wide and rapid dissemination of research output.

More information about this series at http://www.springer.com/series/7092

Roger Lee

Editor

Computer and Information Science

 Springer

Editor
Roger Lee
Software Engineering and Information
  Technology Institute
Central Michigan University
Mt. Pleasant, MI
USA

ISSN 1860-949X   ISSN 1860-9503   (electronic)
Studies in Computational Intelligence
ISBN 978-3-319-86794-6   ISBN 978-3-319-60170-0   (eBook)
DOI 10.1007/978-3-319-60170-0

Printed on acid-free paper

This Springer imprint is published by Springer Nature
The registered company is Springer International Publishing AG
The registered company address is: Gewerbestrasse 11, 6330 Cham, Switzerland

Foreword

The purpose of the 16th IEEE/ACIS International Conference on Computer and Information Science (ICIS 2017) held on May 24–26, 2017 in Wuhan, China, was to together researchers, scientists, engineers, industry practitioners, and students to discuss, encourage, and exchange new ideas, research results, and experiences on all aspects of Applied Computers & Information Technology, and to discuss the practical challenges encountered along the way and the solutions adopted to solve them. The conference organizers have selected the best 17 papers from those papers accepted for presentation at the conference in order to publish them in this volume. The papers were chosen based on review scores submitted by members of the program committee and underwent further rigorous rounds of review.

In Chapter "Big Data and IoT for U-healthcare Security", Mechelle Grace Zaragoza, Haeng-Kon Kim, and Roger Y. Lee present Big Data and IoT for U-healthcare Security architecture to address the challenges in this study. Data privacy of patient and user data is a critical requirement. The architecture will access controls to medical device data.

In Chapter "Retrospection and Perspectives on Pragmatic Software Architecture Design: An Industrial Report", Xiaofeng Cui present a retrospective report on their extensive architecture design of a large-scale mission-critical system conducted in their company. At the end of the project, they considered it as a successful architecting conduct and believed that much experience can be gained from it.

In Chapter "Distributed Coding and Transmission Scheme for Wireless Communication of Satellite Images", Ahmed Hagag, Ibrahim Omara, Souleyman Chaib, Xiaopeng Fan, and Fathi E. Abd El-Samie propose a novel coding and transmission scheme for satellite images broadcasting.

In Chapter "Experimental Evaluation of HoRIM to Improve Business Strategy Models", Yohei Aoki, Hironori Washizaki, Chimaki Shimura, Yuichiro Senzaki, Yoshiaoki Fukazawa discuss aligning organizational goals and strategies is important in business process management (BPM). The Horizontal Relation Identification Method (HoRIM), which is their extension of the GQM+Strategies framework, improves the strategic alignment between organizations.

In Chapter "Combining Lexicon-Based and Learning-Based Methods for Sentiment Analysis for Product Reviews in Vietnamese Language", Son Trinh, Luu Nguyen, Minh Vo propose a framework for sentiment analysis based on combining lexicon-based and learning-based methods for product review sentiment analysis in Vietnamese language.

In Chapter "Reducing Misclassification of True Defects in Defect Classification of Electronic Board", Tokiko Shiina, Yuji Iwahori, Yohei Takada, Boonserm Kijsirikul, and M.K. Bhuyan propose a method to discriminate the defect on the electronic board and the foreign matter attached to the circuit. The purpose of this paper is to reduce the misclassification of the true defect to the pseudo defect in the automatic classification approach.

In Chapter "Virtual Prototyping Platform for Multiprocessor System-on-Chip Hardware/Software Co-design and Co-verification", Arya Wicaksana and Tang Chong Ming describe the implementation of a virtual prototyping platform to address the ever-challenging Multiprocessor System-on-Chip (MPSoC) hardware/software co-design and co-verification requirements.

In Chapter "A Data-Mining Model for Predicting Low Birth Weight with a High AUC", Uzapi Hange, Rajalakshmi Selvaraj, Malatsi Galani, and Keletso Letsholo present a study of data-mining models that predict the actual birth weight, with particular emphasis on achieving a higher Area Under the receiver operating Characteristic (AUC).

In Chapter "A Formal Approach for Maintaining Forest Topologies in Dynamic Networks", Faten Fakhfakh, Mohamed Tounsi, Mohamed Mosbah, Dominique Mery, and Ahmed Hadj Kacem focus on maintaining a forest of spanning trees in dynamic networks. They propose an approach based on two levels for specifying and proving distributed algorithms in a forest.

In Chapter "A Multicriteria Approach for Selecting the Optimal Location of Waste Electrical and Electronic Treatment Plants", Santoso Wibowo and Srimannarayana Grandhi present multicriteria decision-making approach for selecting the optimal location of waste electrical and electronic equipment (WEEE) treatment plants.

In Chapter "Localization Strategy for Island Model Genetic Algorithm to Preserve Population Diversity", Alfian Akbar Gozali and Shigeru Fujimura discuss IMGA, a multipopulation GA model objected to get a better result (aimed to get global optimum) by intrinsically preserve its diversity. They present a new approach which sees an island as a single living environment for its individuals.

In Chapter "HM-AprioriAll Algorithm Improvement Based on Hadoop Environment", Wentian Ji, Qingju Guo, and Yanrui Lei try to improve the efficiency of the mining frequent item sets of AprioriAll algorithm, the Hadoop environment and MapReduce model are introduced to improve AprioriAll algorithm, a new algorithm of mining frequent item sets under the environment of big data HM-AprioriAll algorithm is designed.

In Chapter "Architecture of a Real Time Weather Monitoring System in a Space Time Environment Using Wireless Sensor Networks", Fantazi Walid and Tahar

Ezzedine present a web-mapping framework for collecting, storing, and analysis meteorological data recorded by wireless sensor network.

In Chapter "Mobile Application Development on Domain Analysis and Reuse-Oriented Software (ROS)", Mechelle Grace Zaragoza and Haeng-Kon Kim describe the study of mobile application development with the use of domain analysis. How to reuse software will cause certain advantages and issues upon reviewing this study.

In Chapter "A Transducing System between Hichart and XC on a Visual Software Development Environment", Takaaki Goto, Ryo Nakahata, Tadaaki Kirishima, Kensei Tsuchida, and Takeo Yaku proposed a visual programming development environment for program diagrams called Hichar. In this paper, they describe a visual development environment for the XC language, which is a programming language for XMOS evaluation boards.

In Chapter "Development of an Interface for Volumetric Measurement on a Ground-Glass Opacity Nodule", Weiwei Du, Dandan Yuan, Xiaojie Duan, Jianming Wang, Yanhe Ma, and Hong Zhang have developed an interface to obtain the boundaries of ground-glass opacity (GGO) nodules by using expectation-maximization (EM) algorithm [1] and the histogram method as radiologists' personal habits because the parameters of the EM algorithm and the threshold values of the histogram method can be adjusted.

In Chapter "Efficient Similarity Measurement by the Combination of Distance Algorithms to Identify the Duplication Relativity", Manop Phankokkruad studied the efficient similarity measurement in order to improve the duplication detection techniques in a programming class. This work used the combination of the three proficient algorithms that include Smith-Waterman, Longest Common subsequence, and Damalau-Levenshtein distance to measure the distance between each pair of the code files.

It is our sincere hope that this volume provides stimulation and inspiration, and that it will be used as a foundation for works to come.

May 2017   Guobin Zhu
                                                            Wuhan University, Wuhan, China

                                                                         Shaowen Yao
                                                        Yunnan University, Kunming, China

Reference

1. Miao, Y., Wang, J., Du, W., Ma, Y., Zhang, H.: Volumetric measurement of ground-glass opacity nodules using expectation-maximization algorithm. In: The 4th IIAE International Conference on Intelligent Systems and Image Processing, Kyoto, Japan, pp. 317–321 (2016).

Contents

Contributors

Fathi E. Abd El-Samie Faculty of Electronic Engineering, Department of Electronics and Electrical Communications, Menoufia University, Menouf, Egypt

Yohei Aoki Information and Computer Science, Waseda University, Tokyo, Japan

M.K. Bhuyan Department of Electronics & Electrical Engineering, IIT Guwahati, Guwahati, India

Souleyman Chaib Department of Computer Science, Harbin Institute of Technology, Harbin, China

Xiaofeng Cui Software Engineering Group, BACC, Beijing, China

Weiwei Du Department of Information and Human Science, Kyoto Institute of Technology, Kyoto, Japan

Xiaojie Duan School of Electronics and Information Engineering, Tianjin Polytechnic University, Tianjin, China

Tahar Ezzedine Communication System Laboratory SysCom, National Engineering School of Tunis, University Tunis El Manar, Belvedere, Tunis, Tunisia

Faten Fakhfakh ReDCAD Laboratory, FSEGS, University of Sfax, Sfax, Tunisia

Xiaopeng Fan Department of Computer Science, Harbin Institute of Technology, Harbin, China

Shigeru Fujimura Graduate School of IPS, Waseda University, Kitakyushu, Fukuoka, Japan

Yoshiaoki Fukazawa Information and Computer Science, Waseda University, Tokyo, Japan

Malatsi Galani Department of Computer Science & Information Systems, Botswana International University of Science and Technology, Palapye, Botswana

Takaaki Goto Ryutsu Keizai University, Ibaraki, Japan

Alfian Akbar Gozali Graduate School of IPS, Waseda University, Kitakyushu, Fukuoka, Japan

Srimannarayana Grandhi School of Engineering & Technology, CQUniversity, Melbourne, Australia

Qingju Guo Department of Information Management, Hainan College of Software Technology, Qionghai, People's Republic of China

Ahmed Hagag Department of Computer Science, Harbin Institute of Technology, Harbin, China

Uzapi Hange Department of Computer Science & Information Systems, Botswana International University of Science and Technology, Palapye, Botswana

Yuji Iwahori Graduate School of Engineering, Chubu University, Kasugai, Japan

Wentian Ji Department of Software Engineering, Hainan College of Software Technology, Qionghai, People's Republic of China

Ahmed Hadj Kacem ReDCAD Laboratory, FSEGS, University of Sfax, Sfax, Tunisia

Boonserm Kijsirikul Department of Electronics, Chulalongkorn University, Bangkok, Thailand

Haeng-Kon Kim Catholic University of Daegu, Gyeongsan, South Korea

Tadaaki Kirishima Toyo University, Saitama, Kawagoe-shi, Japan

Roger Y. Lee Central Michigan University, Mount Pleasant, USA

Yanrui Lei Department of Network Engineering, Hainan College of Software Technology, Qionghai, People's Republic of China

Keletso Letsholo Department of Computer Science & Information Systems, Botswana International University of Science and Technology, Palapye, Botswana

Yanhe Ma Tianjin Chest Hospital, Tianjin, China

Tang Chong Ming Department of Electronic Engineering, Universiti Tunku Abdul Rahman, Kampar, Malaysia

Mohamed Mosbah LaBRI, Bordeaux INP, University of Bordeaux, CNRS UMR 5800, Talence, France

Dominique Méry LORIA Laboratory, University of Lorraine, CNRS UMR 7503, Nancy, France

Ryo Nakahata Toyo University, Saitama, Kawagoe-shi, Japan

Luu Nguyen University of Information Technology, Ho Chi Minh City, Vietnam

Ibrahim Omara Department of Computer Science, Harbin Institute of Technology, Harbin, China

Manop Phankokkruad Faculty of Information Technology, King Mongkut's Institute of Technology Ladkrabang, Bangkok, Thailand

Rajalakshmi Selvaraj Department of Computer Science & Information Systems, Botswana International University of Science and Technology, Palapye, Botswana

Yuichiro Senzaki Information and Computer Science, Waseda University, Tokyo, Japan

Tokiko Shiina Graduate School of Engineering, Chubu University, Kasugai, Japan

Chimaki Shimura Information and Computer Science, Waseda University, Tokyo, Japan

Yohei Takada Graduate School of Engineering, Chubu University, Kasugai, Japan

Mohamed Tounsi ReDCAD Laboratory, FSEGS, University of Sfax, Sfax, Tunisia

Son Trinh University of Information Technology, Ho Chi Minh City, Vietnam

Kensei Tsuchida Toyo University, Saitama, Kawagoe-shi, Japan

Minh Vo University of Information Technology, Ho Chi Minh City, Vietnam

Walid Fantazi Communication System Laboratory SysCom, National Engineering School of Tunis, University Tunis El Manar, Belvedere, Tunis, Tunisia

Jianming Wang School of Electronics and Information Engineering, Tianjin Polytechnic University, Tianjin, China

Hironori Washizaki Information and Computer Science, Waseda University, Tokyo, Japan

Santoso Wibowo School of Engineering & Technology, CQUniversity, Melbourne, Australia

Arya Wicaksana Department of Computer Science, Universitas Multimedia Nusantara, Tangerang, Indonesia

Takeo Yaku Nihon University, Tokyo, Setagaya-Ku, Japan

Dandan Yuan School of Electronics and Information Engineering, Tianjin Polytechnic University, Tianjin, China

Mechelle Grace Zaragoza Catholic University of Daegu, Gyeongsan, South Korea

Hong Zhang Tianjin Chest Hospital, Tianjin, China

Big Data and IoT for U-healthcare Security

Mechelle Grace Zaragoza, Haeng-Kon Kim and Roger Y. Lee

Abstract Big Data is a latest topic of interest by many researchers because of its big potential applied in many areas of science and technology. Big Data is by far captivating strong roots in the healthcare ecosystem, but this healthcare data are becoming more complex, which are challenging to solve using common database management tools or simply the traditional data processing application along with the security systems. On the other hand, IoT remainds you to track your health like fitness devices, calorie meters, heart rate monitors, to name a few up to your fridge reminding you that it is basically running out of water." Big Data and IoT are built on networks and cloud computing of gathering data using sensors but challenges using both especially the security for health care is very vital. IoT and Big Data have the potential to transform the way healthcare providers use sophisticated technologies from their clinical and other data repositories and make informed decisions, but without the right security and encryption solution, Big Data and IoT can mean big problems, especially on healthcare security systems. In this study, we discuss the use and application of IoT and Big Data for u-health care. We have presented this architecture to address the mentioned challenges in this study. Data privacy of patient and user data is a critical requirement. The architecture will access controls to medical device data. The patient should be in control of what is being viewed by whom and will allow him/her to view and set the access control policies, maintaining anonymity and masking of data wherever possible.

Keywords U-health care · IoT · Big Data · Security systems

M.G. Zaragoza · H.-K. Kim (✉)
Catholic University of Daegu, Gyeongsan, South Korea
e-mail: hangkon@cu.ac.kr

M.G. Zaragoza
e-mail: mechellezaragoza@gmail.com

R.Y. Lee
Central Michigan University, Mount Pleasant, USA

© Springer International Publishing AG 2018
R. Lee (ed.), *Computer and Information Science*, Studies in Computational
Intelligence 719, DOI 10.1007/978-3-319-60170-0_1

1   Introduction

We live in a communication era where we use the latest technology in everything we do, especially in health care every single day. Numbers of health care-related devices are spreading over to meet the needs of patients or simply for those who are health conscious.

That is one of the main draws of using massive amounts of data, but unfortunately, Big Data and many of the platforms that use it were not designed to address security concerns. If organizations want to ensure their data are secured, they will need to see to it on their own by building those features themselves. If these problems are solved, clinical institutes will be in a better position to truly take advantage of all that Big Data has to offer [1].

2   Background of the Study

As the cost of medical expenses continues to rise, hospitals, clinics, and healthcare institutions can no longer give free hospitalization charges.

The consequence of rising costs increases the penetration of a successful career, which has minimized the provision of unnecessary care, while given the care that is necessary. However, with these even managed situations for patient care, redundant things get to happen, while other effective interventions do not get carried out in part because the number of possible beneficial things to accomplish physicians cannot efficiently keep track of these things. In order to manage this all, an information system must be presented to manage utilization and improve efficiency, quality, and most especially the security of this healthcare system [2].

3   Problem Discussion

IoT is built on networks and cloud computing of gathering data using sensors but challenges using IoT especially the security for health care is very vital. IoT and Big Data have the potential to transform the way healthcare providers use sophisticated technologies from their clinical and other data repositories and make informed decisions, but without the right security and encryption solution, Big Data and IoT can mean big problems especially on healthcare security systems. In this study, we discuss the use and application of IoT and Big Data for u-health care. We have presented this architecture to address the mentioned challenges in this study. Data privacy of patient and user data is a critical requirement. The architecture will access controls to medical device data. The patient should be in control of what is being viewed by whom and will allow him/her to view and set the access control policies, maintaining anonymity and masking of data wherever possible.

Fig. 1 Average cyber-attacks
report from Accenture

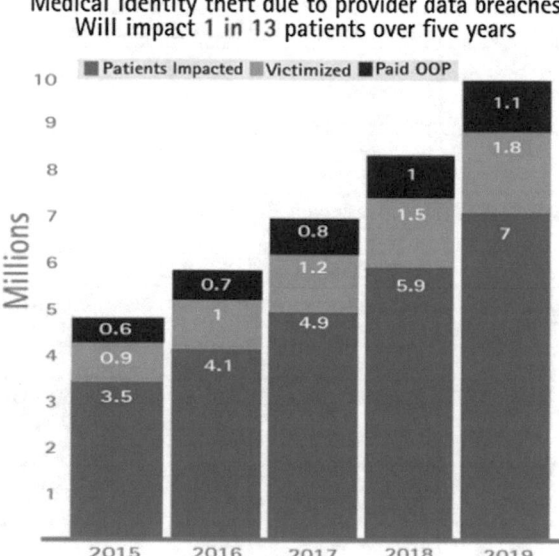

3.1 Security Threats

Figure 1 explains that cyber-attacks in the next 5 years will cost US health systems $305 billion in cumulative lifetime revenue. As Accenture approximates that one in 13 patients—roughly 25 million people—will have personal information, like their social security or financial records, stolen from technology systems in the next 5 years. "What most health systems don't realize is that many patients will suffer personal financial loss as a result of cyber-attacks on medical information," by Kaveh Safavi, CHICAGO; Oct. 14, 2015—[3].

4 Understanding Health Care

In order to understand what health care really is, let us see the history of prior to 1960 of unrelated events rather than a modernized organized effort. Healthcare systems across the globe are focusing on policy in improving the quality of health care delivered. In contrast, let us first see the healthcare quality improvement in earlier time periods from a series of unrelated incidents and its developments.

Table 1 represents the chronological history of health care [4].

Table 1 History of healthcare quality

Chronological summary of key tipping points, individuals by year			
Year(s)	Key tipping points	Key individuals responsible	Country of origin
1854	Quality improvement documentation	Nightingale	England
1861	Sanitary commissions	Barton	USA
1862, 1918	Improvisation and innovation	Pasteur, Blue	France, USA
1879	Sterilization	Chamberland	France
1895, 1956, 1960	Technology	Rontgen, Safar, Laerdal	Germany, USA, France, Norway
1910	Education	Flexner	USA
1881–1955	Pharmaceuticals	Pasteur, von Behring, Kitasato, Descombey, Salk, Kendrick, Eldering Pittman, Fleming	France, Germany Japan, USA, England
1883–1945	Healthcare financing	Bismark, Beveridge, Kaiser	Germany, England, USA
1908	The role of industry and mass production	Ford	USA

4.1 Obama Healthcare Law

4.1.1 Why Do We Need to Care

In November 1, 2016, the enrollment started for this Obama Health Care Reform Bill. This was the former President Barack Obama's reforms in the healthcare industry as this expands the Medicaid and Medicare providing the most affordable insurance to low-to-middle income Americans. The bill is called Obama Care, which includes many provisions that focus on expanding quality, affordable healthcare coverage helping millions of Americans. Obama's plan for healthcare reform outlined what would be in the current Obama healthcare law [5].

But as of now, Americans are already seeing the results: a historic expansion of health care, with costs now growing at the slowest rate in 50 years. Healthcare reform changed the course of history in America. It has been a goal for presidents and progressive leaders since the days of Teddy Roosevelt. Today, finally, no American can be discriminated against because of a preexisting condition. This is how important health care is. As one of the most powerful countries needs improvement to its health care, we believe that we could still improve a lot of things to make health care more beneficial in the community [6].

4.1.2 US to HER Security

As Electronic health record security is concerned, data distribution and its extremely sensitive nature of healthcare information must be protected. The implementation of a very effective security and privacy control must be considered as personal healthcare records are almost digitized. According to guidance issued by the US Department of Health and Human Services in December 2008, the US government is providing incentives to its people to get all healthcare providers using EHR by 2015 in which protected by reasonable administrative, technical, and physical safeguards. It is expected that the adoption of electronic health records will allow up to $100 billion to be saved in healthcare expenditure over a 10-year period, according to research group RAND [7].

5 Internet of Things for Medical Devices (IoT-MD)

The number of increasing sensors used by medical devices, remote, and continuous monitoring of a patient's health is becoming promising. This network of sensors, actuators, and other mobile communication devices, referred to as the Internet of Things for Medical Devices (IoT-MD), is composed to revolutionize the functioning of the healthcare industry.

IoT-MD simply provides a setting in which patient's vital parameters get transmitted by some medical devices through an access point going to a secure cloud-based platform as a storage medium, aggregated and analyzed.

This cloud-based platform is simply the storage of millions of patients and performs analysis in real time, ultimately promoting an evidence-based medicine system [8].

5.1 IoT

How IoT is fast approaching and speeding up its usage in the future?

Figure 2 projects the evolution of interconnected devices in the near future. The existence of IoT is to basically enable all things to be "smart" in a way that things should be connected anytime, anyplace, with anything, and anyone ideally using any path/network and any service. These objects make themselves recognizable to other devices intended to it, and they obtain intelligence by making or enabling context-related decisions. It will serve as a communication device to transmit information about them [9].

As of now, saying that securing properly Internet of Things is a little late but as the IoT increases in diversity and complexity, solutions in resolving these issues

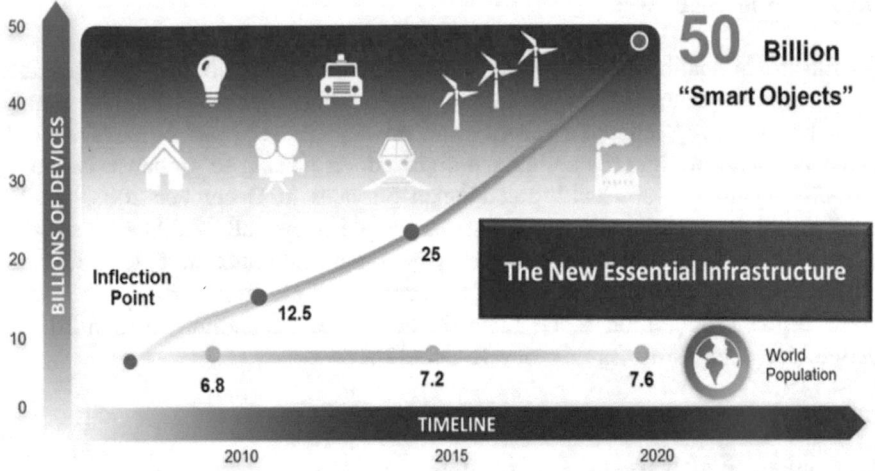

Fig. 2 Interconnected devices and the future of evolution (Cisco 2011)

may not as difficult as we think it is. Technology is growing and progressing that security experts are barely keeping up as it is. The main problem about this increasing number of interconnected devices is mainly the security unless major advances are made to adopt a "security first" idea, as the Internet of Things may always be a risky attempt [10].

6 How Big Is Big Data Is It?

6.1 The 3Vs of Big Data

Big Data has 3 specific areas: Volume, Variety, and Velocity. This is like dealing with massive amounts of data, from all kinds of sources, in all kinds of forms. Traditional BI relies on the so-called structured data which is usually arranged in columns and rows, with every field corresponding to some known source. Big Data is often completely unstructured, as it needs to have a developer to create some kind of mechanism for interpreting the information.

Velocity is defined as the extra capacity you get with distributed processing can also dramatically accelerate computing. Velocity as described in the 3v's of Big Data is the extra capacity you get with distributed processing as velocity is the measure of how fast data is coming in [11].

Figure 3 displays the concept of Big Data was first defined in 2001 by Doug Laney, who specified 3Vs for Big Data: Volume, Velocity, and Variety in his paper titled "3D Data Management: Controlling Data Volume, Velocity and Variety." Shows the 3Vs to describe and explain Big Data in today's era [12] (Figs. 3 and 4).

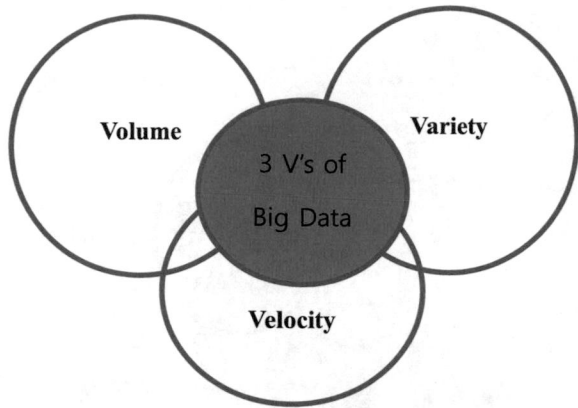

Fig. 3 a Volume: Massive amount of data generated at an unprecedented pace, **b** Velocity: the speed at which the data are created, transformed, stored, analyzed, and visualized, **c** Variety: various types of data in structured, unstructured, and semi-structured formats

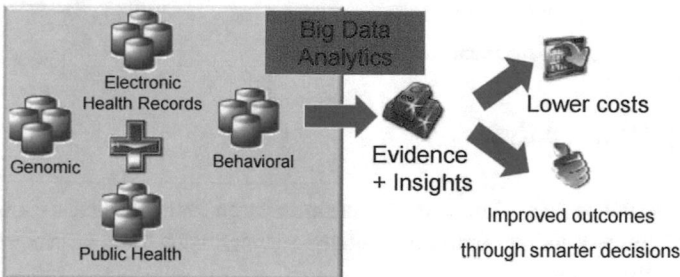

Fig. 4 Overall goals of big data (analytics) in healthcare

In a broader view, it takes advantage of the amounts of data and provides right intervention to the right patient at the right time. It could also be a personalized care to the patient, benefiting the overall components of a healthcare system such as provider, payer, patient, and management [13].

6.2 Challenges and Opportunities in Big Data

Major concern is the privacy issue. We know some computer programs that could readily erase all the details stored and being transported into a larger databases, but stakeholders should be aware about the potential threats these data might become public in the future. Big Data adaptation has the potential to transform health care [14].

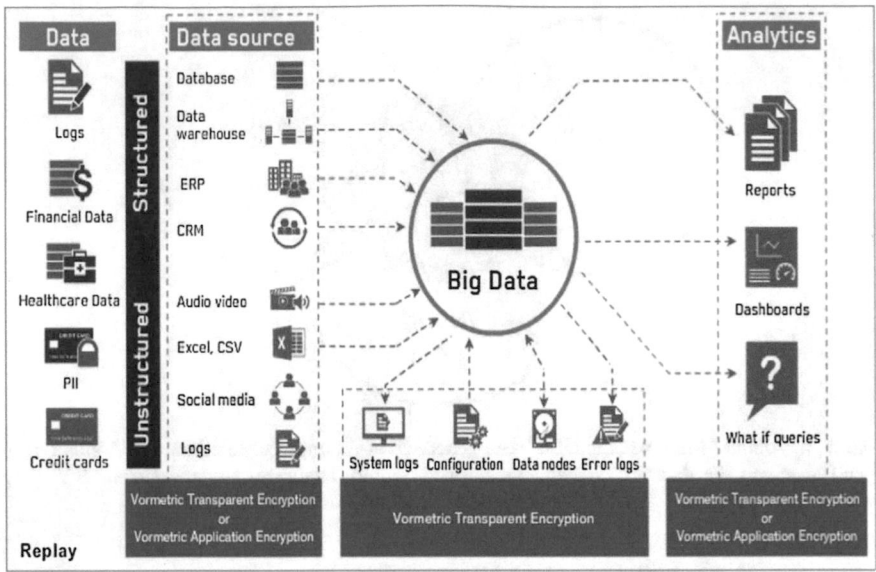

Fig. 5 Securing Big Data environments with vormetric

6.3 Big Data Analytics

As discussed before, Big Data is a collection of large and complex data sets which are hard to process using common database management tools or traditional data processing applications.

As Fig. 5, shows vormetric solutions for Big Data security enable organizations to make use of the benefits of Big Data analytics [15].

7 Discussion

Healthcare data is becoming more complex. Big data analytics have the potential to transform the way healthcare providers use sophisticated technologies to gain insight from their clinical and other data repositories and make informed decisions.

In the future, we will appreciate the widespread implementation and use of Big Data analytics across the healthcare organization, with that several challenges that were mentioned must be addressed. Big Data analytics became too mainstream by this time [16].

Figure 6 shows how a typical IoT hospital revolutionizes as technology and time progress in the field of medicine. Hospitals have started the process of paper-based

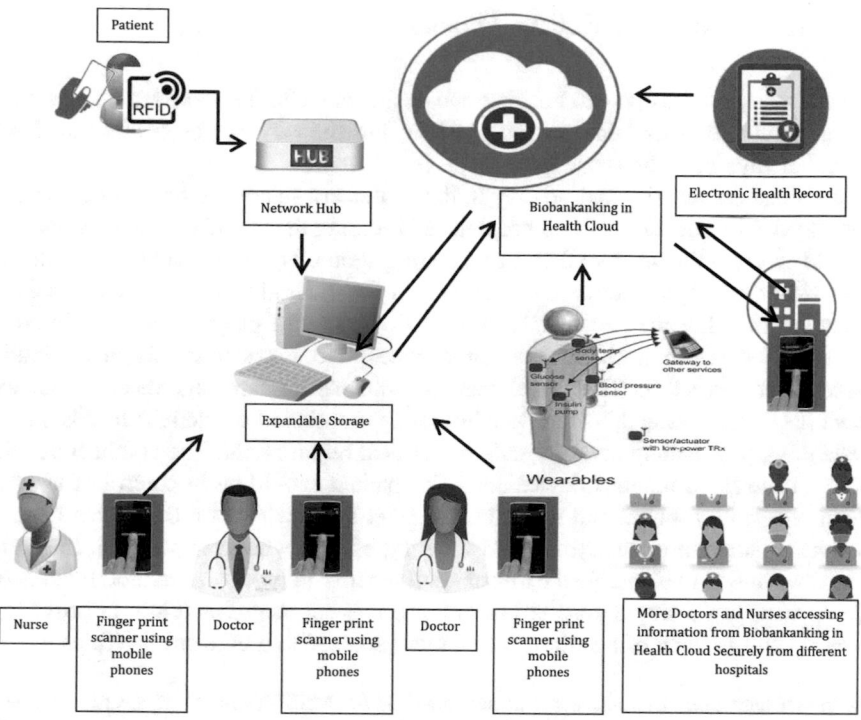

Fig. 6 A proposed secured Internet of Things (IoT) in a single hospital connected to more hospitals in accessing patients' data based on (HIR) Healthcare Informatics Research

medical records to EHR from a traditional writing of a patient's health record. A patient diagnosed with diabetes will have its ID provided to be scanned by the RFID scanner that is connected to the secure biobanking health cloud which stores the (EHR) electronic records of the patient. The mobile app (specific health app for hospital records) will ensure that doctors or the nurses could only see ad retrieve the records stored in biobanking. To do his, the program will have to ensure the doctor's data are already stored in the program as well as the time inputted and in the near future, the records being collected in a specific field provided for example (patient name, birth date, or diagnosis). This proposed architecture is possible, though IoT and Big Data. This will ensure that the most sensitive data being disseminated are safe. That those who can only access are the authorized persons listed on the system.

This may seem very basic, as we try to expand the cloud capacity to store the needed information to be distributed, this will create more advances and privacy-wise useful in the field of medicine [17].

8 Conclusion and Future Works

Health care paved its way to be more successful as technology proved its existence to make our lives as laid-back as possible, but the advantages depict that there would always be a disadvantages.

Information security and privacy in the healthcare sector are both critical issues they need to be addressed. As to keeping and securing the records of the patients, one should consider the other goal of u-heathcare systems. That is to not only provide the best treatment a patient should deserve, but the overall treatment of the patient including their hospital records. To solve these issues, the proposed secured Internet of Things (IoT) in a single hospital connected to more hospitals in accessing patients' data based on (HIR) Healthcare Informatics Research and its attributes, we have presented this architecture to address the mentioned challenges in this study. Data privacy of patient and user data is a critical requirement. The architecture will access controls to medical device data. The patient should be in control of what is being viewed by whom and will allow him/her to view and set the access control policies, maintaining anonymity and masking of data wherever possible. In future work, we intend to elaborate the information security (a proposed secured Internet of Things (IoT) in a single hospital) in order to make it useful as a tool in providing security and privacy of health data to a more secured and efficient health care.

Acknowledgements This research was supported by the MSIP (Ministry of Science, ICT and Future Planning), Korea, under the C-ITRC (Convergence Information Technology Research Center) support program (IITP-2016-H8601-15-1007) supervised by the IITP (Institute for Information & Communication Technology Promotion).

This research was supported by the International Research & Development Program of the National Research Foundation of Korea (NRF) funded by the Ministry of Science, ICT, and Future Planning (Grant Number: K 2014075112).

References

1. Buckley, J.: 7 Big Data Security Concerns. Qubole, 7 May 2015
2. Bates, D.W., Kuperman, G.J., Wang, S., Gandhi, T., Kittler, A., Volk, L., Spurr, C., Khorasani, R., Tanasijevic, M., Middleton, B.: Ten commandments for effective clinical decision support: making the practice of evidence-based medicine a reality. J. Am. Med. Inf. Assoc. **10**(6), 523–530 (2003)
3. Cyberattacks Will Cost U.S. Health Systems $305 Billion Over Five Years, Accenture Forecasts, 14 Oct 2015
4. Sheingold, B.H., Hahn, J.A.: The history of healthcare quality: the first 100 years 1860–1960. Int. J. Afr. Nurs. Sci. **1**, 18–22 (2014)
5. ObamaCare Facts: Facts on the Affordable Care Act. http://obamacarefacts.com/obama-health-care-bill/
6. Health care reform is helping millions of Americans get better health care. That's worth fighting for. https://www.barackobama.com/obamacare/
7. Howarth, F.: Security challenges in the US healthcare sector, Bloor Research, A white paper, December 2010

8. Khanna, A., Misra, P.: The Internet of Things for Medical Devices—Prospects, Challenges and the Way Forward, A white paper
9. Vermesan, O., Friess, P.: Internet of Things: Converging Technologies for Smart Environments and Integrated Ecosystems
10. Ar. Amster: Internet of Things: Big Data and Data Security Problems, 27 Apr 2016
11. Junk, D.: Big Data vs Traditional Approaches to Enterprise Reporting. http://blog.apterainc. com/business-intelligence/big-data-vs-traditional-approaches-to-enterprise-reporting. Accessed 18 Sept 2015
12. Shan, T.: Big Data 3Vs, Leading by Game-Changing Cloud, Big Data and IoT Innovations, 10 Oct 2013
13. Sun, J., Reddy, C.K.: Big Data Analytics for Healthcare. https://www.siam.org/meetings/ sdm13/sun.pdf
14. Kayyali, B., Knott, D., Kuiken, S.V.: The big-data revolution in US health care: accelerating value and innovation
15. BIG DATA SECURITY Vormetric Data Security Use Cases. https://www.vormetric.com/ data-security-solutions/use-cases/big-data-security
16. Raghupathi, W., Raghupathi, V.: Big data analytics in healthcare: promise and potential. J. Health Inf. Sci. Syst.
17. Dimitrov, D.V.: Medical internet of things and big data in healthcare. Health Inf. Res. 22(3), 156–163 (2016)

Retrospection and Perspectives on Pragmatic Software Architecture Design: An Industrial Report

Xiaofeng Cui

Abstract It is commonly recognized that the research on software architecture has enjoyed a golden age of innovation and concept formulation, and began to enter the mature stage of utilization. It is a natural expectation at present that the relevant concepts and methods have been populated and the improvements have been achieved in practice. In this paper, we give a retrospective report on our extensive architecture design of a large scale mission-critical system conducted in our company. The project has been trying to incorporate newly proposed concepts and methods from academic realm, and keeping introspective during the whole process. At the end of the project, we considered it as a successful architecting conduct, and believed that much experience can be gained from it. We describe the process, approaches, and results of this industrial practice. Experience and perspectives from the practitioners' points of view are analyzed and summarized. The contribution of this paper is twofold. Firstly, it presents a real-world software architecture design phenomenon, shares first-hand experience and perspectives on the pragmatic architecture design practices, therefore provides some insights into the state of the practice. Secondly, it may inspire further research on more effective and efficient methods and tools for better practices.

Keywords Software architecture design · Architectural significant requirements (ASR) · Architecture design decision (ADD) · Architecture design rationale

1 Introduction

Software architecture has emerged as a distinct discipline for about three decades [1], and enjoyed a golden age of innovation for more than 10 years [2]. Clements and Shaw [3] pointed out in 2006 that "*software architecture has been considered an unexceptional, essential part of software system building; the research in this*

X. Cui (✉)
Software Engineering Group, BACC, Beijing, China
e-mail: cuixf@pku.org.cn

© Springer International Publishing AG 2018
R. Lee (ed.), *Computer and Information Science*, Studies in Computational Intelligence 719, DOI 10.1007/978-3-319-60170-0_2

field has enjoyed a golden age of innovation and concept formulation, and began to enter the more mature stage of quiet discipline and unremarkable utilization."

In this situation, it is a natural expectation at present that the architecture-relevant concepts and methods have been populated, and the improvements have been achieved in practice. Much research however, indicates the gap between the state of the art and the state of the practice. As Moore et al. [4] pointed out earlier, *"solving a problem in theory and in practice are very different."* In order to investigate the state of the practice of software architecture, there have emerged many empirical studies [5–7]. Most of them are based on surveys and drawing conclusions from interview feedbacks. The extensive first-hand reports from industrial practitioners are lack.

In this paper, we give a retrospective report on our extensive architecture design of a large scale mission-critical system conducted in our company for about 2 years. The project has been trying to incorporate newly proposed concepts and methods from academic realm, and keeping introspective during the whole process. At the end of the project, we considered it as a successful architecting conduct, and believed that much experience can be gained from it. Therefore we think it is worthy to summarize the experience on our interested problems about software architecture.

Based on our experience of industrial practice, we are interested in a series of questions to explore the pragmatic application of academic concepts and methods. Because of the space limitation, we concentrate on two of them in this paper:

- Q1: When, where, how and by whom are Architectural Significant Requirements (ASRs) identified? What are the main difficulties with the identification of ASRs in practice?
- Q2: How are Architecture Design Decisions (ADDs) made in practice? Are they made by intuition or formal procedures? What are the main difficulties with the architecture decision making in practice?

In this paper, we describe the process, approaches, and results of this industrial practice. Experience and perspectives from the practitioners' points of view are analyzed and summarized. This report tries to present a real-world software architecture design phenomenon, share first-hand experience and perspectives on the pragmatic architecture design practices, and therefore provide some insights into the state of the practice. It also aims to inspire further research on more effective and efficient methods and tools for better practices.

The rest of this paper is organized as follows. Section 2 presents related work. Section 3 gives an overview of our architecting project. Section 4 describes in detail the requirements elicitation and architecture design deliberation of the project. Section 5 gives the retrospective analysis, including perspectives on the research problems and suggestions on better methods. Finally, Sect. 6 presents concluding remarks and future work.

2 Related Work

2.1 Architectural Significant Requirements

The inevitable intertwining of requirements and design has long been concerned by both academic and practice realms extensively [8]. At the architecture level, it is commonly conceived that some requirements, especially quality requirements (QR, or non-functional requirements, NFR), have crucial impacts on software architecture. These requirements are commonly called Architectural Significant Requirements (ASRs). Quite a lot of research has been conducted on the elicitation, identification, specification, and evaluation of ASRs [9].

Capilla et al. [9] highlight the importance and the role of quality requirements for architecting and building complex software systems. They point out that architecture challenges arise because the required quality attributes are informally stated during requirements elicitation and analysis. Niu et al. [10] argue that despite the increasing effort in engineering enterprise systems' requirements, little is known about the analysis of architecture interactions and tradeoffs. They propose a framework consisting of an integrated set of activities to help requirements analysis in practice.

While most research's focus is on NFRs, e.g., Niu et al. [10], Anish et al. [11] are interested in those Functional Requirements (FRs) that have significant impact on architecture, i.e. ASFR. They conduct a qualitative interview study aimed at identifying ASFRs' categories from various business domains. They also indicate that architects intuitively recognize ASRs and often seek out relevant stakeholders in order to ask Probing Questions (PQs) that help them acquire the information they need.

To systematically characterize ASRs, Chen et al. [12] present an evidence-based framework that consists of four sets of characteristics: Definition, Descriptive, Indicators, and Heuristics. They characterize ASRs as having wide impact on the system, exhibiting tradeoffs with other requirements, introducing constraints, breaking assumptions, and often being difficult to achieve.

Cleland-Huang [13] introduces a novel approach that uses "architecturally savvy personas" to depict quality concerns as personalized, architecturally significant user stories, and then leveraging these user stories to drive a system's initial design. The personas they created were intended to help reason about ASRs and to drive a series of architectural decisions.

2.2 Architecture Design Decisions

A series of methods have been proposed for the design or derivation of software architecture, e.g., Attribute-Driven Design (ADD) [14], Siemens Four-Views (S4V) [15], etc. Hofmeister et al. [16] compare five industrial software architecture design methods and extract a general model, which classifies the kinds of design activities into architectural analysis, synthesis, and evaluation. Architectural analysis articulates ASRs. Architectural synthesis results in candidate solutions. Architectural evaluation ensures that the architectural decisions are the right ones.

The software architecture community has paid attention to design decisions and rationale for a long time. Hofmeister et al. [16] present a model of software architecture that consists of three components: elements, form, and rationale. Bosch [17] promotes that design decisions should be represented as first-class entities in software architectures. Tyree and Akerman [18] claim that a key to demystifying architecture products lies in the architecture decisions concept. Kruchten et al. [19] propose an ontology of software design decisions. Jansen et al. [20] develop a tool to model the software architecture as the composition of design decisions. Tang et al. [21] introduce the rationale-based model to support design rationale capture and traversal.

Falessi et al. [22] make a comparative survey on the decision-making techniques for software architecture design. They point out that the software engineering literature describes several techniques to choose among architectural alternatives, but it gives no clear guidance on which technique is more suitable than another, and in which circumstances. Their work represents a first attempt to reason on meta-decision-making, i.e., the issue of deciding how to decide.

Focusing on how architects make the transition from requirements to architecture, Hebisch et al. [23] argue that studies on the decision-making process examine the architecting process only after the architects have already established different architectural alternatives. They propose an approach for structuring the problem first in order to understand how appropriate solutions might look like, and creating architectural alternatives more directly from the requirements, enabling the evaluation of design decisions even before an architecture exists.

A lot of efforts have been made to cope with the interplay of requirements and architecture and their synergetic development [24, 25]. Chen [26] aims at advancing Requirements Engineering (RE) practice through the co-development of requirements and architecture by utilizing the relationship between ASRs and ADDs. They argue that in real software projects, requirements engineering is seldom separated from architectural design. However, the co-development of requirements and architecture is not well supported by current techniques and methods.

2.3   Empirical Study on Architecture

Kruchten and Stafford [1] pointed out in 2006 that *"The importance of software architecture for software development is widely recognized, yet transfer of innovative techniques and methods from research to practice is slow."* In recent years, many empirical studies have been made on the state of the practice of software architecture.

Heesch and Avgeriou [5] give a report on findings of a survey that they have conducted with 53 industrial software architects to find out how they reason in real projects. The results are interpreted with respect to the industrial context and the architecture literature. They find that the greatest part of the participants does not seem to follow one particular architecture approach from the literature; instead they at least partially adopt architecture activities to define their own customized approach to architecture.

Anvaari et al. [6] conduct an exploratory study in Norwegian electricity industry, and point out that regarding the architectural decision-making process, most of the companies are not using well-known academic approaches, they are rather using their own procedures. They also show that the relationships among the actors of a software ecosystem could significantly affect the architectural-decision making process.

Dan et al. [7] point out that much research exists on architectural decisions, but little work describes architectural decisions in the real-world. They present the results of a survey with 43 architects from industry. They study characteristics of 86 real-world architectural decisions and indicate that dependencies between decisions and the effort required to analyze decisions are major factors that contribute to their difficulty. Good architectural decisions tend to include more decision alternatives than bad decisions. Finally, they found that 86% of architectural decisions are group decisions.

Daneva et al. [27] point out that only recently has RE research yielded the first five empirical studies of software architects' perceptions of QRs. To determine a possible range of views on how architects cope with QRs, they interviewed 20 software architects and found that software architects feel it's important to gain a deep understanding of the QRs and use this to deliver good architecture design.

Although there are such empirical studies, most of them are based on surveys and draw conclusions from interview feedbacks. This paper gives a detailed description and extensive analysis of a real-world architecting work conducted by ourselves, intending to provide some insights into the state of the practice, and inspire some points for further improvement.

3 Project Overview

3.1 Project Background

The project we study in this paper is an upgrading project of a large-scale aerospace flight control software system. The existing system has been used for about 8 years and carried out a series of missions successfully. The objective of the upgrading is to improve the system in terms of capability, usability, as well as new functionalities, so that it can serve well for the upcoming missions which will be much more complex and challenging.

In order to achieve the objective, our company set up an architecting team in 2013. The team comprised 11 experienced engineers. All of them are familiar with the existing system, and most of them have taken part in the development of the existing system. Because the system is quite large (6 subsystems, 21 configuration items, about 2000KLOC for the existing system; 11 subsystems, 40 configuration items, anticipated 3000KLOC for the new system), and the objective of this upgrading is ambitious, our company decided to grant the team 2 years (part-time job) to accomplish this architecting work.

By the end of 2015, the team finished its work and submitted the final architecture design of the new system. According to the review of the design and the feedback from the following development work in 2016, the whole process of architecting is considered to be successful and the design results are considered to be satisfactory.

3.2   Process and Methods

From the beginning of the architecting work, we have been trying to incorporate new concepts and methods promoted by the software architecture research community, and keep the architecting work effective, efficient, and pragmatic at the same time. The whole process of architecting is the synergetic deliberation of requirements and architecture solutions. For the sake of clarity, we roughly divide all the activities into 3 parts, as shown in Fig. 1 (which by no means implies a waterfall process of 3 stages). Table 1 summarizes the main approaches and points adopted in each part of the architecting, which will be explained in more detail in the next section. As mentioned in the introduction section, we concentrate on the "early part" of architecting in this paper, while omit other activities such as architecture documentation, evaluation, etc.

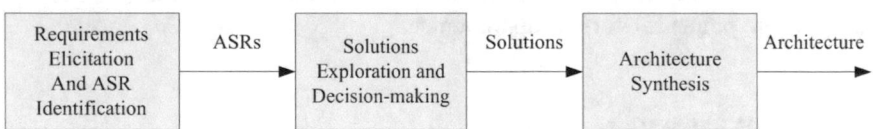

Fig. 1 Three parts of architecture design activities

Table 1 Approaches adopted in each part of the architecture design

Architecture design activities	Approaches and main points adopted
Requirements elicitation and ASR identification	Improvement points analysis of existing system; Requirements investigation of future missions
Solutions exploration and decision-making	Group deliberation; Group decision-making; Goal-oriented analysis
Architecture synthesis	Structure composition; Global design problems resolution; Key design decisions articulation

4 Architecting Process Description

4.1 Architectural Requirements Elicitation

According to literature and our experience, architects' thorough understanding of requirements is crucial to the success of architecting. Therefore in this project, we didn't split our team into requirement engineers and architecture designers. Our architecting team is at the same time requirement analysis team.

Based on the motivation of this upgrading project, we firstly defined 6 goals as the high-level goals of new system, which are 3 "high" (high performance, high reliability, high availability) and 3 "easy" (easy build, easy verify, easy use). Although various business functionalities (e.g., data processing, commanding and control, etc.) and other quality attributes (e.g., security) are important to the system as well, our analysis shows that the aforementioned 6 goals are crucial to the ultimate objective of new system. Table 2 lists and explains these goals in detail.

Then we made clear two main sources of the requirements for this architecting work: the improvement points of existing system, and the key demands of the future missions. In other words, we understood that the ultimate objective of our architecting work is to establish an architecture that can support the dramatic improvement of the existing system, as well as satisfy the new requirements of the foreseeable future missions.

In order to find out the improvement points of existing system, we conducted a thorough drawback analysis, a survey of the improvement appeals from its operators, and a deep statistical analysis of problem reports during its service history. In order to understand the demands of future missions, we made an investigation on their key characteristics. Although the information of future missions cannot be acquired comprehensively yet, the mission profiles can act as good inputs for the

Table 2 Design goals of the new system

Goals	Explanation
High performance	The future missions demand much higher performance requirements on the new system, including time performance (real-time) and space capability (throughput)
High reliability	The new system needs to have higher reliability than the existing system, so that it can satisfy the demand of critical missions
High availability	The new system needs to work correctly for longer time than the existing system, and support online maintenance
Easy to build	The new system needs to be built more easily than the existing system, which means requiring fewer human efforts and less time
Easy to verify	The new system needs to be verified easily, which means the verification cost can be decreased and quality can be guaranteed more easily than the existing system
Easy to use	The new system needs to have higher usability. The user experience can be increased and the operation efficiency can be increased dramatically

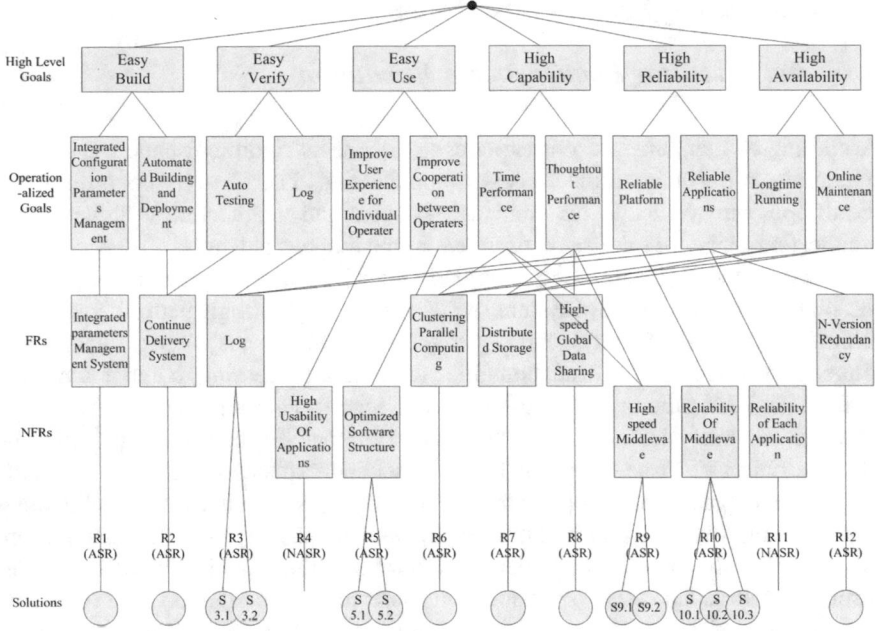

Fig. 2 The tree of goals, requirements, and solutions

architecture design, e.g., the highly volume data throughout, the long-lasting mission duration, etc.

When we had found out the improvement points and new requirements, we established functional requirements and non-functional requirements of the new system. All the requirements are linked to the high-level goals of the new system. The intermedium level contains operationalized goals, which bridge the requirements and high-level goals. The two-level goals and the requirements are depicted in Fig. 2 as a tree diagram (a simplified version for brevity). This diagram demonstrates the supporting relationships between the goals and requirements at different levels.

The last step of requirements analysis is to identify which requirements are architectural significant. We identified ASRs by the following criteria: if one requirement can be resolved within one configuration item of the system, or if they can be implemented at the detailed design or coding stages, then they are not architectural significant (i.e., NASR). On the contrary, if one requirement leads to a global design, or its design impacts more than one configuration items in the system, then it is an ASR. Figure 2 also depicts the results of our ASR identification, where 10 requirements are identified as ASRs, and 2 are identified as NASRs.

4.2 Architecture Solutions Deliberation

As the ASRs were identified, our next work was to resolve them. Although there are many literature-proposed methods for step-wise design solution exploration and formal decision making, we prefer to accomplish these jobs based on team expertise and group deliberation. Based on our experience in many projects, we think the latter approaches are more efficient and pragmatic.

We found two practical problems in this work and tackled them with distinct principles. Firstly, how can we resolve the requirements with their interrelationships in concern? Our principle is: if the ASRs are independent to each other, they can be resolved independently; if two or more ASRs are interdependent, then we combine them together and resolve them as one ASR. Secondly, how many solutions do we need to explore for one requirement? Our principle is: if we can find one solution and the design group can agree on it as an ideal solution, then we stop exploring more other solutions; if we cannot reach a consensus on the solutions of one requirement, which means we have not yet find the ideal solution, then we explore more solutions and make decisions on the explored solutions.

Considering the ASRs in Fig. 2, we found that the solutions to R1 and R2 are to newly build an Integrated Parameters Management subsystem and a Continue Delivery subsystem. We did not think it is necessary to explored more solutions for R1 and R2. The requirements R6, R7, R8, R12 are likewise. On the contrary, for the requirements R3, R5, R9, and R10, we found various design options and then explored more than one solutions for them (including the existing solutions). We explain these design and decisions as well as rationale in more detail as follows.

- Example 1: Requirement R3 and its solutions S3.1 and S3.2

The goal of R3 is to provide a log service with high capability, so that applications in the system can log a large amount of running information used for status monitoring, exception detecting, and root cause analysis. We explored two solutions for this requirement: S3.1 and S3.2, as shown in Fig. 3. Solution S3.1 is the existing solution of log service, where all log messages are sent to a remote monitor directly. This solution works well if the amount of log messages is not very large. But if we want the applications to report as much as information, the network traffic can increase dramatically and may block the running of applications, which is not tolerable for real-time applications. For this reason, we explored a new solution S3.2, where the log messages are sent to a local agent. The agent can buffer the log messages and send them to the monitoring and storage sites. This solution supports large amount of log messages because the agent can buffer the messages so that the applications need not worry about delay. In addition, the logs are stored in a dedicated site, so that they can be mined afterwards. Because S3.2 can satisfy the goal of R3 in a more effective way than S3.1, we can make a decision selecting S3.2.

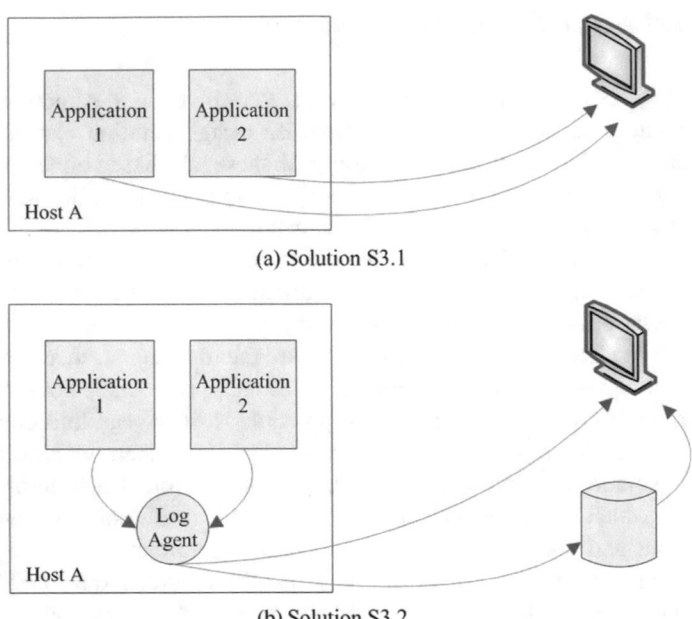

(a) Solution S3.1

(b) Solution S3.2

Fig. 3 Solutions to the requirement R3

- Example 2: Requirements R5 and its solutions S5.1 and S5.2

The goal of R5 is to improve the structure of existing system so that it has higher cohesion and looser coupling characteristics, and improve the efficiency of human operation. There are more than one improve points for this purpose. We pick up one of them as an example here. Figure 4 depicts two solutions for this purpose: S5.1 and S5.2. Solution S5.1 is the existing design of the planning and scheduling functionalities in the system, where the Planning software and Scheduling software belong to two different sub-systems. The Planning software outputs the plan files, which are sent to the Scheduling software as inputs. The Scheduling software schedules the applications in the system according to the plans. Based on this design, two operators are needed, one carrying out the planning operation and another carrying out the scheduling operation. Because the goal of R5 is to improve the operating efficiency, we explored a new solution S5.2, where the Planning software and Scheduling software are put into one new subsystem, i.e. the Planning and Scheduling subsystem. Then the interface between the two functionalities becomes an internal interface, which is easier to maintenance. In addition, only one operator is needed to operate the subsystem. Therefore we can make the design decision selecting the solution S5.2.

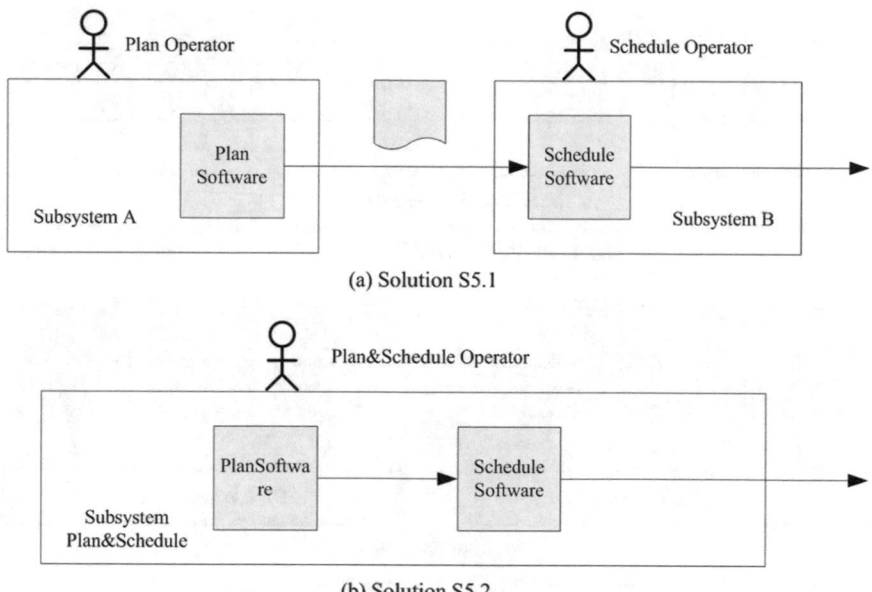

(a) Solution S5.1

(b) Solution S5.2

Fig. 4 Solutions to the requirement R5

- Example 3: Requirements R10 and its solutions S10.1, S10.2 and S10.3

The goal of R10 is to improve the reliability of the whole system. As the system needs to carry out multiple missions at the same time, we use the "Separation Policy" to improve reliability. There are different solutions to implement this requirement. We explored three of them: S10.1, S10.2, and S10.3, as shown in Fig. 5. S10.1 depicts one solution where one middleware instance supports multiple missions at the same time. This solution is straightforward, but has two drawbacks. The first one is that the middleware needs to suffer high volume of messages of multiple missions. The second one is that if one mission has an exception and make the middleware fail, e.g., blocked, then the other mission will be impacted. Therefore this solution is not a highly efficient and reliable solution. The solution S10.2 provide each mission one instance of the middleware, so that the fault in one mission cannot impact the other missions. Moreover, the burden of each middleware instance is decreased. This is a highly reliable solution. However, S10.2 cannot support inter-mission interactions which are necessary in some circumstances. Then we explored the solution S10.3, where multiple instances of middleware are established, and each application connects to one or more middleware instances based on their communication needs. This solution have the advantages of both S10.1 and S10.2, therefore it is the selected solution for R10.

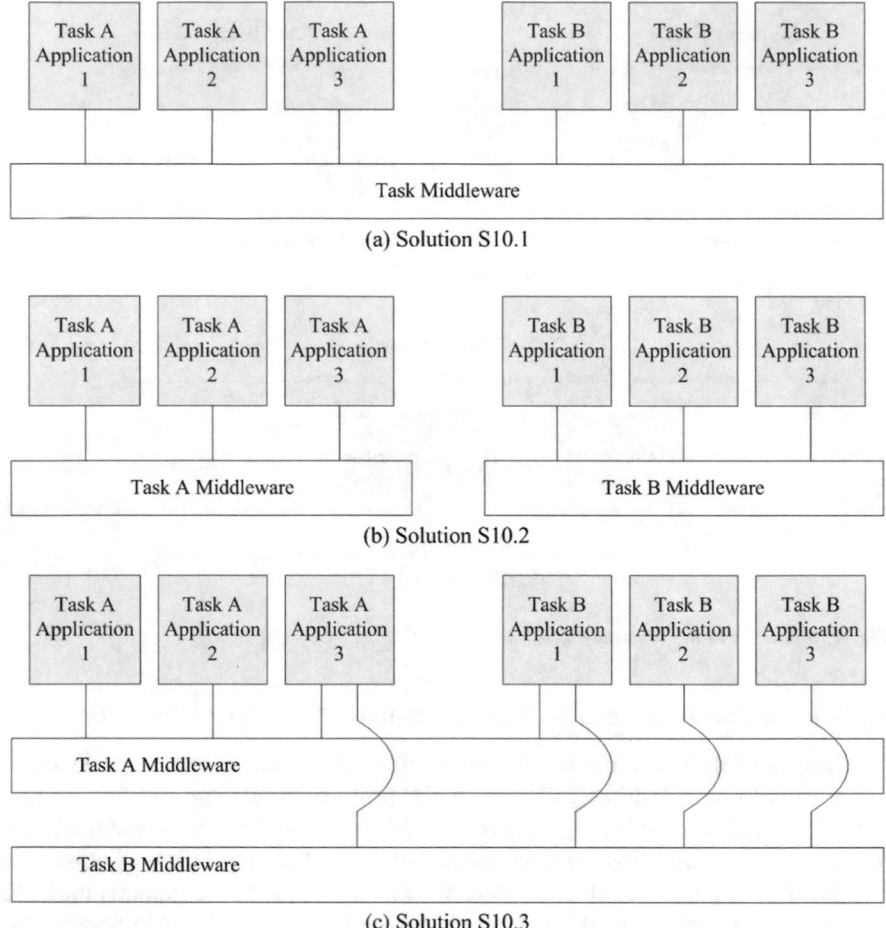

Fig. 5 Solutions to the requirement R10

4.3 Architecture Synthesis

The solutions for the ASRs are solution segments of the architecture. The final part of architecting work is to synthesize the whole architecture. In our opinion, the architecture of the system consists of: (1) the top one or two levels of the division of system (subsystem and configuration items), (2) the interfaces among these divisions, and (3) the global design decisions (ASRs and their solutions, decisions and rationale). For the sake of brevity, we omit the articulation of whole architecture design results here.

5 Retrospective Analysis

5.1 Perspectives on the Research Questions

Based on our experience on this project (as well as other projects we have carried out), we sum up our perspectives on the questions mentioned in Sect. 1. Table 3 lists the questions as well as both perspectives from literature and our practice.

For question Q1, our experience shows that architects in practice do not worry about the identification of ASRs indeed. We think a good architecture need to be designed upon the consideration of all FRs and NFRs, and only architects rather than requirements providers can figure out what requirements need to be considered at architecture level, while others can be solved at detailed design or coding level. Therefore ASRs are the outputs of architecting, rather than the inputs of it. It is commonly considered that the communication of ASRs from requirements field to architecture field is one major obstacle to architecture design. We think the

Table 3 Questions and perspectives

Questions	Perspectives from literature	Perspectives from our practice
Q1. When, where, how and by whom are Architectural Significant Requirements (ASRs) identified? What are the main difficulties with the identification of ASRs in practice?	ASRs are identified in the requirements field, and passed to architecture design field by stakeholders. The identification and communication are usually obstacles to architecture design "*Stakeholders and requirements engineers frequently fail to express or effectively communicate ASRs to the architects, preventing the architects from making informed design decisions.*" [12]	ASR identification is naturally an integral part of the architects' job, because only the architects know what is architectural relevant. The pragmatic architects do not feel the difficulty of ASR identification. They instead think about all the perceived FRs and NFRs, and find out key issues need to resolve at architecture level
Q2. How are Architecture Design Decisions (ADDs) made in practice? Are they made by intuition or formal procedures? What are the main difficulties with the architecture decision making in practice?	ADDs are expected to be made by a systematic and formal methodology. The experience-based decisions are supposed to be irrational decisions "*Architects usually rely on their intuition and experience to create and choose between architecture alternatives. The viability of a decision can be checked by evaluating the architecture only after it has been created.*" [22]	For the sake of efficiency, the formal methods are not easy to be embraced in practice. But practical design activities can benefit from the decision-driven methods. These methods make architects explore the solution space thoroughly and make comparison explicitly, which can clearly improve the quality of design results

identification of ASRs is the responsibility of architects whereas not requirements engineers. In this situation, the communication of ASRs is not a real problem in practice.

For question Q2, our main point is that the consensus-based group deliberation is the effective and efficient design decision-making paradigm in practice. It is more pragmatic than formal or automated decision-making under most circumstance, because work efficiency and subjective consensus are critical factors of success in real-world projects. In our experience, a good design needs to satisfy (at least nearly) all the design participants. If one participant strongly resists one solution, there must be some problems with it. If one group cannot select a solution that satisfies all participants, they'd better explore more new solutions. On the contrary, when using formal decision-making methods, the result may be just the candidate with the highest score, which doesn't necessarily mean it can satisfy the group well.

5.2   What We Suggest for Better Methods and Tools

In our experience, there are still needs for effective architecting methods and tools in practice. In our opinion, many existing methods are not ideal because they have not concerned difficulties in real-world projects, e.g., the labor or time cost of the method, the importance of subjective approval, etc. We highlight two suggestions for better methods and tools in practice.

- Recording of the design process, decision, and rationale

During the process of our design and decision-making, we found that the real difficulty of large architecting projects is not the solution exploring or decision-making, because these jobs depend on the expertise of designers, and good designers should have ability to carry out these tasks well. The real difficulty, however, is the lack of efficient means to record the thoughts, arguments, rationale in the whole design processes. For large projects, the number of design and decision points can be several dozens and even more than one hundred (while only very few of them are mentioned as examples in this paper). During the deliberation process, it is not easy to record these kinds of information totally and precisely, because people think and talk faster than write. Even if we can record everything with electronic recorders, it is even not easy to find desired information from them because there is such long time duration and we cannot search it with keywords or other clue. In academic case studies or demo examples, all kinds of information can be recorded and described delicately. But for real-world projects, especially large ones, the schedule is tight and human resources are valuable, how to tackle the recording difficulty is a practical problem.

- Analysis of the architectural relevance of software defects

As aforementioned, in order to find out the drawbacks of the existing system, we considered the problem reports of existing system during its service history as invaluable assets to inspire improvements. But we found it is not easy to analyze the architectural relevance of some problems. Is the problem implying an architecture drawback, or is it only a lower-level design or implementation defect? The problems need to be analyzed one by one. For large volume of reports, it is a high burden of human efforts. If there are systematic methods for the problem reporters (usually users, programmers, etc.) to identify one problem's architectural relevance correctly in the first place, the conduct of architecture upgrading will be much more effective and efficient.

6 Conclusions and Future Work

The software architecture realm has enjoyed its golden age and is expecting blooming and exciting application effects. We have conducted a series of software architecting work in our company, trying to incorporate new concepts, principles, and methods to improve the architecting results. We make a respective report on a large-scale and extensive architecting project, elaborating the process and methods involved, comparing perspectives from literature and practice, and presenting our suggestions on the future methods for more effective and pragmatic application.

The report is not a thorough empirical study, and the result only represents one kind of circumstance, but we think it can contribute some ideas for further research and practice. Because of the space limitation, this paper only concentrates on some aspects of architecting, i.e., ASRs and ADDs. We plan to embrace more methods in the future architecting practices, e.g., architecture documentation and evaluation, and make more thorough analysis on the practice experience.

References

1. Kruchten, P., Stafford, J.: The past, present, and future for software architecture. IEEE Softw. **23**, 22–30 (2006)
2. Clements, P., Shaw, M.: The golden age of software architecture revisited. IEEE Softw. **26**, 70–72 (2009)
3. Shaw, M., Clements, P.: The golden age of software architecture. IEEE Softw. **23**, 31–39 (2006)
4. Moore, M., Kazman, R., Klein, M., Asundi, J.: Quantifying the value of architecture design decisions: lessons from the field. Int. Conf. Softw. Eng. 557–562 (2003)
5. Heesch, U.V., Avgeriou, P.: Mature architecting—a survey about the reasoning process of professional architects. In: Ninth Working IEEE/IFIP Conference on Software Architecture, pp. 260–269 (2011)
6. Anvaari, M., Conradi, R., Jaccheri, L.: Architectural decision making in enterprises: preliminary findings from an exploratory study in Norwegian electricity industry. Software Architecture, pp. 162–175 (2013)
7. Dan, T., Galster, M., Avgeriou, P.: Difficulty of architectural decisions—a survey with professional architects. Eur. Conf. Softw. Archit. 192–199 (2013)

8. Swartout, W., Balzer, R.: On the inevitable intertwining of specification and implementation. Commun. ACM **25**, 438–440 (1982)
9. Capilla, R., Babar, M.A., Pastor, O.: Quality requirements engineering for systems and software architecting: methods, approaches, and tools. Requirements Eng. **17**, 255–258 (2012)
10. Niu, N., Xu, L.D., Cheng, J.R.C., Niu, Z.: Analysis of architecturally significant requirements for enterprise systems. IEEE Syst. J. **8**, 850–857 (2014)
11. Anish, P.R., Daneva, M., Clelandhuang, J., Wieringa, R.J., Ghaisas, S.: What you ask is what you get: understanding architecturally significant functional requirements. Requirements Eng. Conf. 86–95 (2015)
12. Chen, L., Alibabar, M., Nuseibeh, B.: Characterizing architecturally significant requirements. IEEE Softw. **30**, 38–45 (2013)
13. Cleland-Huang, J.: Meet Elaine: a persona-driven approach to exploring architecturally significant requirements. IEEE Softw. **30**, 18–21 (2013)
14. Bass, L., Clements, P., Kazman, R.: Software Architecture in Practice, 3rd edn. Pearson (2013)
15. Hofmeister, C., Nord, R., Soni, D.: Applied Software Architecture. Addison-Wesley (2000)
16. Hofmeister, C., Kruchten, P., Nord, R.L., Obbink, H., Ran, A., America, P.: A general model of software architecture design derived from five industrial approaches. J. Syst. Softw. **80**, 106–126 (2007)
17. Bosch, J.: Software architecture: the next step. In: 1st European Workshop on Software Architecture (EWSA'04), pp. 194–199 (2004)
18. Tyree, J., Akerman, A.: Architecture decisions: demystifying architecture. IEEE Softw. **22**, 19–27 (2005)
19. Kruchten, P.: An ontology of architectural design decisions in software-intensive systems. In: 2nd Groningen Workshop on Software Variability Management, pp. 54–61 (2004)
20. Jansen, A., Jan, V.D.V., Avgeriou, P., Hammer D.K.: Tool support for architectural decisions. In: IEEE/IFIP Conference on Software Architecture, vol. 4 (2007)
21. Tang, A., Jin, Y., Han, J.: A rationale-based architecture model for design traceability and reasoning. J. Syst. Softw. **80**, 918–934 (2007)
22. Falessi, D., Cantone, G., Kazman, R., Kruchten, P.: Decision-making techniques for software architecture design: a comparative survey. ACM Comput. Surv. 43, 88–87 (2011)
23. Hebisch, E., Book, M., Gruhn, V.: Scenario-based architecting with architecture trace diagrams. In: IEEE/ACM 5th International Workshop on the Twin Peaks of Requirements and Architecture (TwinPeaks), pp. 16–19 (2015)
24. Cleland-Huang, J., Hanmer, R.S., Supakkul, S., Mirakhorli, M.: The twin peaks of requirements and architecture. IEEE Softw. **30**, 24–29 (2013)
25. Galster, M., Mirakhorli, M., Cleland-Huang, J., Franch, X., Burge, J.E., Roshandel, R.: Towards bridging the twin peaks of requirements and architecture. ACM Sigsoft Softw. Eng. Notes **39**, 30–31 (2014)
26. Chen, F.: From architecture to requirements: relating requirements and architecture for better requirements engineering. Requirements Eng. Conf. 451–455 (2014)
27. Daneva, M., Herrmann, A., Buglione, L.: Coping with quality requirements in large, contract-based projects. IEEE Softw. **32**, 84–91 (2015)

Distributed Coding and Transmission Scheme for Wireless Communication of Satellite Images

Ahmed Hagag, Ibrahim Omara, Souleyman Chaib, Xiaopeng Fan and Fathi E. Abd El-Samie

Abstract In this chapter, we propose a novel coding and transmission scheme for satellite images broadcasting. First, we use a 2D-DWT to divide full-size satellite image band into four sub-bands; three of them are the details and a small version of image band in LL sub-band. Second, our scheme utilizes coset coding based on distributed source coding (DSC) for the LL sub-band to achieve high compression efficiency and a low encoding complexity. After that, without syndrome coding, the transmission power is directly allocated to band details and coset values according to their distributions and magnitudes without forward error correction (FEC). Finally, these data are transformed by Hadamard matrix and transmitted over a dense constellation. Experiments on satellite images demonstrate that the proposed scheme improve the average image quality by 2.28 dB, 2.83 dB and 3.66 dB over LineCast, SoftCast-3D and SoftCast-2D, respectively, and it achieves up to 6.26 dB gain over JPEG2000 with FEC.

Keywords Wireless communication · Satellite images · DSC · LineCast · SoftCast

A. Hagag (✉) · I. Omara · S. Chaib · X. Fan
Department of Computer Science, Harbin Institute of Technology,
Harbin 150001, China
e-mail: ahagag88@gmail.com

I. Omara
e-mail: i_omara84@yahoo.com

S. Chaib
e-mail: chaib@hit.edu.cn

X. Fan
e-mail: fxp@hit.edu.cn

F.E. Abd El-Samie
Faculty of Electronic Engineering, Department of Electronics and Electrical
Communications, Menoufia University, Menouf 32952, Egypt
e-mail: fathi_sayed@yahoo.com

© Springer International Publishing AG 2018
R. Lee (ed.), *Computer and Information Science*, Studies in Computational
Intelligence 719, DOI 10.1007/978-3-319-60170-0_3

1   Introduction

Today, satellite images have been widely demonstrated in several applications and remote sensing projects, ranging from independent land mapping services to government and military activities [1–5]. Some of these applications and projects are desperately needed the use of wireless communication to transmit the satellite images. Therefore, satellite wireless systems have become an important topic needs to study. In agriculture, the need for observational data, aircraft and satellite remote sensing is playing an expanded role in farm management with the huge volume of information represented by the satellite images. Also for the meteorologists, the satellite images play an essential role to explore water vapor, clouds properties, aerosols and absorbing gases.

Recently, global positioning systems (GPS) [6] used the satellite images to provide the location and more information for the users, anywhere on the Earth. Google has announced that they can track ships at sea in real time through satellites and will provide this tracking as a public service on the Internet [3]. Now, wireless communication tools able to transfer large packets of data quickly and efficiently over distances. In the wireless communication, there are existed more and more technologies and techniques used. However, in this chapter, we interested in the broadcast satellite images communication over the wireless channel as shown in Fig. 1.

In [7] and [8], Shannon has concluded that there are two main challenges to transmit data over the wireless channel: source coding (data compression) and channel coding (forward error correction (FEC) and modulation scheme). Source coding is designed in separation from the channel coding, and the problems can be solved and the source can be transmitted without loss of information if the channel is point-to-point (i.e., unicast communication) and the channel quality is known or can be easily measured at the source, by selection for the optimal transmission rate for the channel and the corresponding FEC and modulation [8]. However, for broadcast/multicast, challenges are still founded, where each receiver observes a different channel quality.

Fig. 1 Satellite images broadcasting

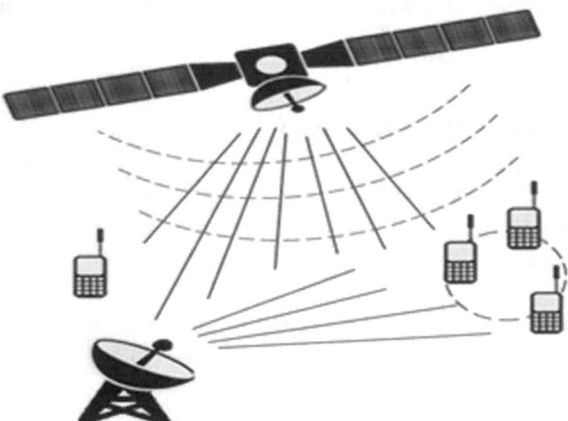

In contrast to the previous separate design, and the following approaches presented in [9–11], we proposed a new distributed coding and transmission scheme for broadcasting satellite images to a large number of receivers. In our scheme, we avoided the annoying cliff effect founded in the digital broadcasting schemes by using a linear transform between the transmitted image signal and the original pixels luminance. We apply distributed source coding (DSC) based on the Slepian–Wolf theory [12] technique in satellite image multicast to achieve high efficient compression and low encoding complexity. The experimental results, conducted on the selected datasets and, show the superior of the proposed scheme compared with the state-of-the-art methods.

The rest of the chapter is organized as follows: Sect. 2 provides the related works that commonly used. Section 3 introduces the proposed scheme including both encoder and decoder. The evaluation environment and the experimental results are presented in Sect. 4. Section 5 provides the conclusion of the chapter.

2 Related Works

2.1 Digital Broadcasting

Satellite broadcast systems separate source coding and channel coding based on Shannon's source-channel separation theorem [7, 8].

In the source code, there are many techniques recently used to compress the satellite images [13], one of them that used in this chapter is JPEG2000 [14]. The standard is based on the wavelet coding. JPEG2000 supports the 9/7 and the 5/3 integer wavelet transforms. After transformation, coefficient quantization is adapted to individual scales, and sub-bands and quantized coefficients are arithmetically coded.

In the channel coding, FEC and digital modulation are used in the digital scheme. However, we cannot accommodate all channel conditions simultaneously for all receivers in a broadcast scenario, such the transmission rate has to adapt to actual channel condition by adjusting the channel coding rate and modulation.

2.2 SoftCast

SoftCast [9, 10] is one of analog approaches that design within a joint source-channel coding (JSCC) framework. These schemes can optimize received distortion by a linear combination of original signals, which are then directly transmitted over a dense constellation. SoftCast transmits the linear transform of the image/video signal directly in analog channel without quantization, entropy coding and FEC. This principle naturally enables a transmitter to satisfy multiple receivers with diverse channel qualities. The SoftCast encoder consists of the following steps:

transform, power allocation and direct dense modulation. In the transform phase, there are two versions for SoftCast: SoftCast-2D [9] and SoftCast-3D [10]. In the SoftCast-2D, a two-dimensional discrete cosine transform (2D-DCT) is used to remove the spatial redundancy of an image/video frame. In the SoftCast-3D, a three-dimensional DCT (3D-DCT) is used to remove spatial and spectral redundancies of a group of images/video frames. In the second step, power allocation minimizes the total distortion by optimally scaling the transform coefficients. After that, SoftCast employs a linear Hadamard transform to make packets with equal power and equal importance. Finally, the data are directly mapped into the wireless symbols by a very dense QAM.

At the receiver, SoftCast uses a linear least square estimator (LLSE) as the opposite operation of power allocation and Hadamard transform. Once the decoder has obtained the DCT components, it can reconstruct the original image/video frames by taking the inverse of the transform.

2.3 LineCast

In [15], Wu et al. have proposed a new coding and transmission scheme, called LineCast, for broadcasting satellite images to a large number of receivers.

The LineCast encoder consists of the following steps: read image line by line, transform, power allocation and transmission. In the first step, the LineCast reads image line by line. In the transform step, decorrelate each line by a one-dimensional DCT (1D-DCT). After that, a scalar modulo quantization is performed on the DCT components with the same technique used in [11] and [16], which partitions source space into several cosets and transmits only coset indices to the decoder. After the scalar modulo quantization, a power allocation technique is employed to these frequencies, which are scaled with different parameters at different frequencies to minimize the total distortion between them. Finally, as SoftCast, LineCast employs a linear Hadamard transform to get resilience against packet losses. Then, the data are directly mapped into the wireless symbols by a very dense 64 K-QAM without digital FEC and modulation.

At the receiver, LineCast uses LLSE to reconstruct the line signal from the DCT components. Moreover, the side information is generated to aid the recovery of transform coefficients in scalar modulo dequantization. Finally, the minimum mean square error (MMSE) is used to denoising the reconstructed signal.

3 Proposed Scheme

The proposed scheme transmits a satellite image over the raw OFDM channel directly without FEC and digital modulation. Figure 2 shows the framework of our proposed scheme.

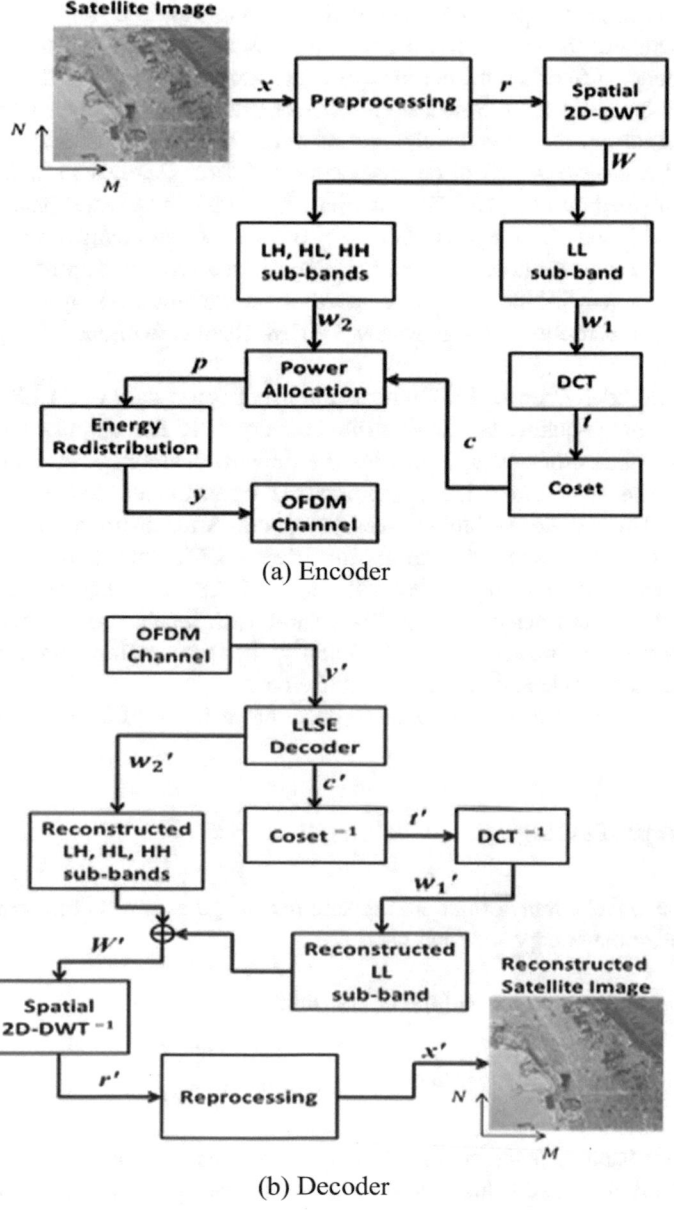

(a) Encoder

(b) Decoder

Fig. 2 Framework of our proposed scheme

At the encoder, the proposed scheme first proposes to do a simple preprocessing by subtraction of the integer mean value for each band x, then scaled to reduce the band's energy. After that, the compression and transmission of the first band in our scheme is the same as in SoftCast, which consists of DCT, power allocation and Hadamard transform. Whereas, the rest of image bands r's are firstly decomposed by spatial two-dimensional discrete wavelet transform (2D-DWT), and the results of W are divided into LL, LH, HL and HH sub-bands. The LL sub-band is encoded using DSC-based. The cosets values c of the LL sub-band and other three sub-bands are power-allocated. Finally, the proposed scheme employs a Hadamard transform to redistribute energy as used in communication systems and then transmitted the results y through raw OFDM channel without FEC and digital modulation.

At the decoder, after the PHY returns the list of coded data y', a LLSE is applied to provide a high-quality estimate of the recovery data. The output consists of two groups. The first group is w_2', contains the three reconstructed sub-bands LH, HL and HH. The other group is c', contains the reconstructed cosets values of LL sub-band. The decoder technique used the previous reconstructed band to reconstruct the LL coset values t', then applies inverse DCT transform on coset reconstruction and gets a reconstructed sub-band LL in w_1'. Once the decoder has obtained the reconstructed the four DWT sub-bands, it can reconstruct the original band data r' by the inverse 2D-DWT. Finally, the reprocessing data are applied to reconstruct the whole satellite image band in x'.

In the following subsections, we will describe each part of the proposed scheme.

3.1 Preprocessing

Before the wavelet transformation, the satellite image band x_i is first preprocessing to normalize the energy for each band i.

$$r_i = (x_i - m_i)/s_i, \text{ and}$$

$$m_i = |\text{mean}(x_i)|_{\text{Round}}, s_i = \sqrt{\frac{\|x_i\|^2}{NM}}$$

where each band x_i contains N rows and M columns. For each band x_i, the mean value m_i and the scaled value s_i need to be transmitted to the receiver as a metadata, so that x_i can well recovered.

3.2 Transformation

As in SoftCast [9], excepting the first image band that is compression and transmission, which consists of DCT, power allocation and Hadamard transform. Each band of the reminder bands in the satellite image is decomposed firstly by spatial 2D-DWT with level 1 decomposition to obtain four sub-bands in W_i that contains w_1 and w_2; three of these sub-bands (LH, HL and HH) represent the details of the image band in w_2, and the fourth sub-band LL in w_1. After the preprocessing step, we apply a 1D-KLT to remove redundant information within a spectral dimension. After that, we adopt a 2D-DCT for every band as a spatial decorrelator.

$$W_i = \mathrm{DWT}(r_i)$$

The Daubechies 9/7 filter bank, introduced in [17], is used for transformation.

3.3 Coset Encoding for LL Sub-band

Coset coding is a typical technique used in DSC. For the image band, the coset values represent the details of image band. We already have the details of the full-size image band in the step of DWT in w_2 and still have a small version of image band in w_1. In the proposed scheme, we use a coset coding proposed in [18] to only the LL sub-band for every image band. Coset coding achieves the satisfied compression because the coset value has typically lower entropy than the source value; moreover, the DSC provides a low encoding complexity than traditional compression techniques.

In the proposed scheme, we apply a DCT for the LL sub-band of band i and encode the DCT coefficients t_i to get coset values c_i.

$$c_i = t_i - \left\lfloor \frac{t_i}{q_i} + \frac{1}{2} \right\rfloor q_i$$

where q_i is the coset step for band i that calculated by estimating the noise of the decoder prediction as shown in [16].

3.4 Power Allocation and Transmission

The coset data c_i and the detail sub-bands in w_2 for band i are scaled for optimal power allocation in terms of minimizing the distortion [9] with scaling factor g_i.

$$g_i = \lambda_i^{-1/4}\left(\sqrt{\frac{P}{\sum_{i=1}^{F}\sqrt{\lambda_i}}}\right)$$

where λ_i is the variance of c_i and w_2, F is the number of frequencies and P is the total transmission power. The signal p_i after power allocation is

$$p_i = g_i \cdot (c_i, w_2)$$

To redistribute energy, we protect the weighted signal p_i against packet losses with multiplying it by a Hadamard matrix H. The outputs then transmitted using the raw OFDM without FEC and modulation steps.

$$y_i = H \cdot p_i.$$

In addition to the image data, the encoder sends a small amount of metadata contains m_i, s_i and λ_i to assist the decoder in inverting the received signal. The metadata are transmitted in the traditional way (i.e., OFDM in 802.11 PHY with FEC and modulation). Here, a BPSK is used for modulation, and half-rate convolutional code is used for FEC to protect the metadata from channel errors.

3.5 LLSE Decoder

At the receiver, for each transmitted signal y_i, we receive a noisy signal

$$y_i' = n_i + y_i$$

where n_i is a random channel noise for band i. The LLSE is used to reconstruct the original signals as follows:

$$(c_i', w_2') = \Lambda_{q_i} R_i^T \left(R_i \Lambda_{q_i} R_i^T + \Lambda_{n_i}\right)^{-1} y_i'$$

where $R_i = H \cdot g_i$ for each band i, Λ_{n_i} is the covariance matrix of n_i and $\Lambda_{(c_i', w_2')}$ is a diagonal matrix whose diagonal elements are the variances, λ_i, of the individual band.

3.6 Inverse Coset and Transforms

After the recovery for signal, we reconstruct the DCT coefficients for LL sub-band in t_i' by coset decoding in [18]

$$t'_i = c'_i - \left\lfloor \frac{c'_i}{q'_i} + \frac{1}{2} \right\rfloor q'_i$$

where q'_i is the coset step for each band i that calculated based on the noise and the previous received band as shown in [18]. After that, we recover the group w_1' by applying the inverse DCT (i.e., the reconstructed LL sub-band for band i).

In the final step, we combine the two groups w_1' and w_2' into one dataset W' and apply the inverse 2D-DWT to get r_i', then reprocess it and get the final image band. Note that m_i and s_i are transmitted as metadata by the encoder.

$$r'_i = \text{IDWT}(W'_i)$$
$$x'_i = (r'_i s_i) + m_i$$

4   Experimental Results

In experiments, we evaluate the performance of the proposed scheme under broadcasting channels. We compare the proposed scheme with some well-known and recent works (SoftCast-2D [9], SoftCast-3D [10], LineCast [15] and JPEG2000 [14] with different combinations of FEC rates and modulation methods).

In the SoftCast-2D [9], we apply the 2D-DCT for every band separately in the selected satellite images. Whereas, the SoftCast-3D [10] uses 3D-DCT to remove both spectral and spatial redundancies from the satellite image data. Different from SoftCast, LineCast compresses every scanned line of an image by the transform-domain scalar modulo quantization without prediction. We have implemented the LineCast [15] for every band separately in the selected satellite images.

In the JPEG2000, DWT and principal components analysis (PCA) are often used as spectral decorrelators [19]. In our experiments, JPEG2000 with PCA in the spectral decorrelator is used in the digital source coding. The satellite image data are first decorrelated with PCA, and then, a 2D-DWT with 5 levels decomposition is used as spatial decorrelating. In the DWT, the 9/7 filter has been used, and the reference software BOI [20] is used for implementation.

The proposed scheme has been tested on several satellite images (see Fig. 3). The selected images are from the aerials volume of the USC-SIPI image database [21]. The aerials images with a resolution of $1024 \times 1024 \times 3$ are represented by 8 bits per pixel per band (bpppb). All images are publicly available for download [19].

Quality is computed comparing the original image $x(M, N, \lambda)$ with the recovered image $x'(M, N, \lambda)$. We evaluate these methods using the peak signal-to-noise ratio (PSNR), a standard metric of image/video quality.

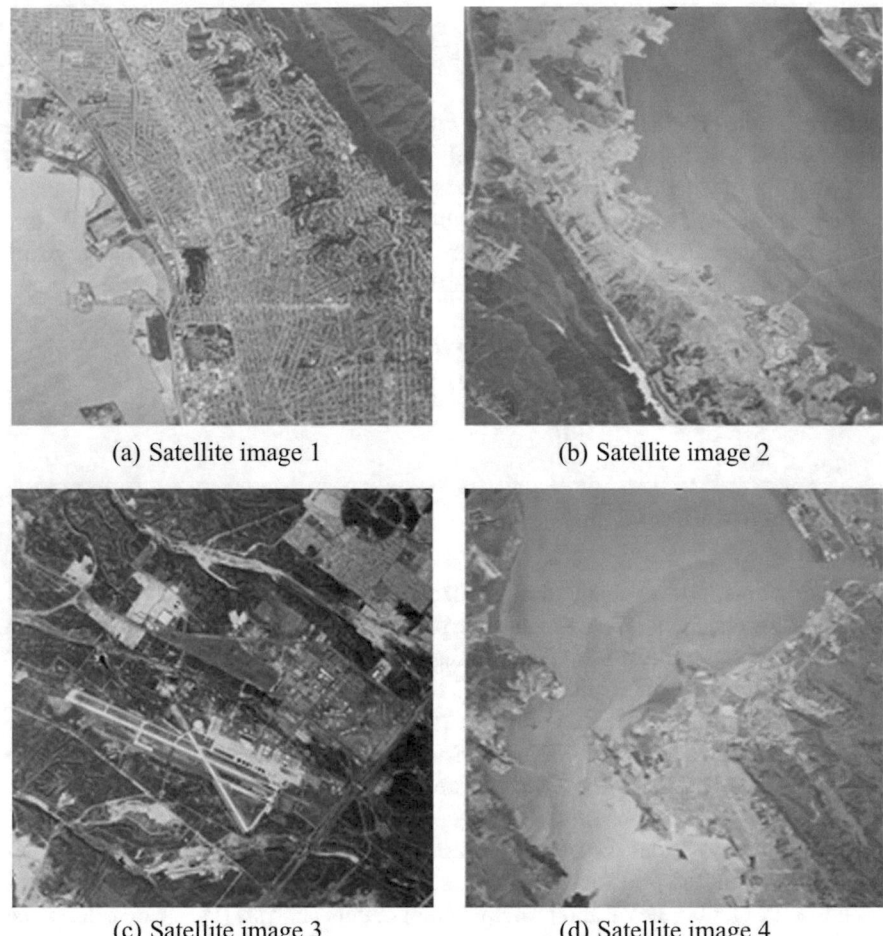

(a) Satellite image 1   (b) Satellite image 2

(c) Satellite image 3   (d) Satellite image 4

Fig. 3 Test satellite images

$$PSNR_{(dB)} = 10 \cdot \log_{10}\left(\frac{L^2}{MSE}\right) (dB)$$

$$MSE = \frac{1}{M \cdot N \cdot \lambda} \sum_{i=1}^{M} \sum_{j=1}^{N} \sum_{k=1}^{\lambda} \left[x(i,j,k) - x'(i,j,k)\right]^2$$

where λ is the number of satellite image bands, and L is the maximum possible pixel value of the satellite image (i.e., $L = 2^B - 1$, where B is the bit depth). Typical values for the PSNR in lossy satellite image compression are between 30 and 50 decibels (dB), provided the bit depth is 8 bits, where higher is better.

The reconstructed satellite images PSNR of each scheme under different channel SNR between 5 and 25 dB are given in Fig. 4. We can see that our proposed

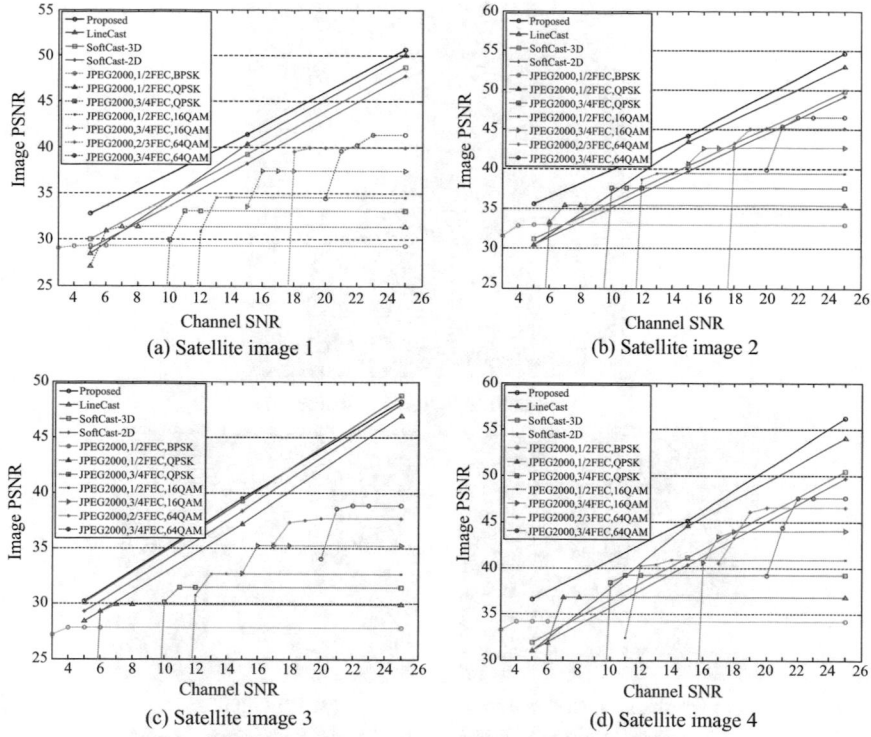

Fig. 4 Broadcasting performance comparison

scheme achieves better performance than other schemes in broadcasting satellite images. It is clear that the separable source-channel coding (i.e., JPEG2000 with FEC and modulation) exhibits cliff effect for all seven conventional transmission approaches. In contrast, the proposed scheme avoids the annoying cliff effect (i.e., when the channel SNR increases, the reconstruction PSNR increases accordingly, and vice versa).

The visual quality comparison is given in Fig. 5. The channel SNR is set to be 5 dB. Satellite image 1 is selected for comparison. The proposed scheme shows better visual quality than others. Our proposed scheme delivers a PSNR gain up to 2.28 dB, 2.83 dB, 3.66 dB and 6.26 dB over LineCast, SoftCast-3D, SoftCast-2D and JPEG2000 with FEC and modulation, respectively.

(a) Satellite image 1 full size.

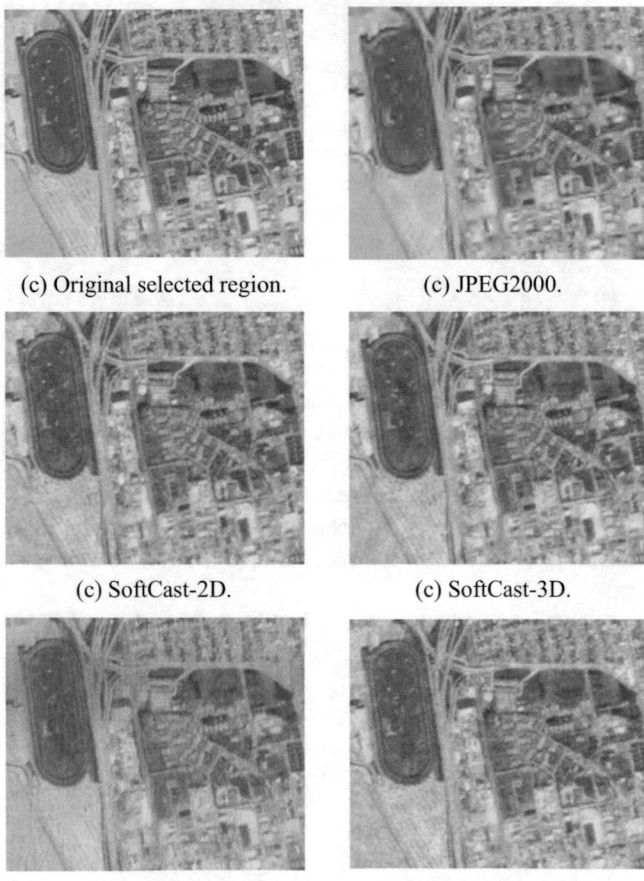

(c) Original selected region.   (c) JPEG2000.

(c) SoftCast-2D.   (c) SoftCast-3D.

(c) LineCast.   (c) Proposed scheme.

Fig. 5 Visual quality comparison at SNR = 5 dB

5 Conclusion

In this chapter, we propose a novel scheme for satellite image broadcasting. The proposed scheme takes advantage of the spectral redundancies in satellite image by the use of DSC principle into spectral bands. We utilize characteristics of the wavelet transform to represent the details of the image band. Furthermore, we also use coset coding to achieve low encoding complexity and efficient compression performance. The proposed scheme avoided the annoying cliff effect founded in the digital broadcasting schemes by using a linear transform between the transmitted image signal and the original pixels' luminance. Experimental results demonstrate that the proposed scheme achieved better performance than other state-of-the-art schemes.

References

1. Crowley, M.D., Chen, W., Sukalac, E.J., Sun, X., Coronado, P.L., Zhang, G.Q.: Visualization of remote hyperspectral image data using Google Earth. In: IEEE International Conference on Geoscience and Remote Sensing Symposium, pp. 907–910 (2006)
2. Evans, B., Werner, M., Lutz, E., Bousquet, M., Corazza, G.E., Maral, G.: Integration of satellite and terrestrial systems in future multimedia communications. IEEE Wirel. Commun. **12**(5), 72–80 (2005)
3. Google Can Track Ships At Sea: Including US Navy; Detailed Maps Planned of Sea Bottom. http://defense.aol.com/2012/05/17/google-satellites-can-track-every-ship-at-sea-including-us-na/ (2012)
4. Zhong, Z., Pi, D.: Forecasting satellite attitude volatility using support vector regression with particle swarm optimization. IAENG Int. J. Comput. Sci. **41**(3), 153–162 (2014)
5. Medjahed, S.A., Saadi, T.A., Benyettou, A., Ouali, M.: Binary cuckoo search algorithm for band selection in hyperspectral image classification. IAENG Int. J. Comput. Sci. **42**(3), 183–191 (2015)
6. Kaplan, E.D., Hegarty, C.J.: Understanding GPS: Principles and Applications. Artech House, Norwood, MA, USA (1996)
7. Shannon, C.E.: A mathematical theory of communications. Bell Syst. Tech. J. **27**, 379–423 (1948)
8. Shannon, C.E.: Two-way communication channels. In: The 4th Berkeley Symposium on Mathematical Statistics and Probability (1961)
9. Jakubczak, S., Rahul, H., Katabi, D.: One-size-fits-all wireless video. In: Proceedings of the Eighth ACM SIGCOMM HotNets Workshop, New York City, NY (2009)
10. Jakubczak, S., Katabi, D.: A cross-layer design for scalable mobile video. In: Proceedings of the 17th Annual International Conference on Mobile Computing and Networking (2011)
11. Fan, X., Wu, F., Zhao, D., Au, O.C.: Distributed wireless visual communication with power distortion optimization. IEEE Trans. Circuits Syst. Video Technol. **23**(6), 1040–1053 (2013)
12. Slepian, D., Wolf, J.: Noiseless coding of correlated information sources. IEEE Trans. Inf. Theory **19**(4), 471–480 (1973)
13. Blanes, I., Magli, E., Serra-Sagristà, J.: A tutorial on image compression for optical space imaging systems. IEEE Geosci. Remote Sens. Mag. **2**(3), 8–26 (2014)
14. Taubman, D., Marcellin, M.: JPEG2000: image compression fundamentals, standards, and practice. In: International Series in Engineering and Computer Science. Kluwer, Norwell, MA (2002)

15. Wu, F., Peng, X., Xu, J.: LineCast: line-based distributed coding and transmission for broadcasting satellite images. IEEE Trans. Image Process. **23**(3), 1015–1027 (2014)
16. Fan, X., Wu, F., Zhao, D., Au, O. C., Gao, W.: Distributed soft video broadcast (DCAST) with explicit motion. In: Data Compression Conference (DCC), pp. 199–208 (2012)
17. Antonini, M., Barlaud, M., Mathieu, P., Daubechies I.: Image coding using wavelet transform. IEEE Trans. Image Process. **1**(2), 205–220 (1992)
18. Fan, X., Wu, F., Zhao, D.: D-Cast: DSC based soft mobile video broadcast. In ACM International Conference on Mobile and Ubiquitous Multimedia (MUM), Beijing, China (2011)
19. Penna, B., Tillo, T., Magli, E., Olmo, G.: Transform coding techniques for lossy hyperspectral data compression. IEEE Trans. Geosci. Remote Sens. **45**(5), 1408–1421 (2007)
20. Aulí-Llinàs, F.: BOI codec. http://www.deic.uab.cat/~francesc/software/boi/ (2014)
21. The USC-SIPI Image Database. http://sipi.usc.edu/database/ (1977)

Experimental Evaluation of HoRIM to Improve Business Strategy Models

Yohei Aoki, Hironori Washizaki, Chimaki Shimura, Yuichiro Senzaki and Yoshiaoki Fukazawa

Abstract Aligning organizational goals and strategies is important in Business Process Management (BPM). The Horizontal Relation Identification Method (HoRIM), which is our extension of the GQM+Strategies framework, improves the strategic alignment between organizations. GQM+Strategies aligns the strategies across organizational units at different levels by a strategy model, which is a tree structure of strategies called a GQM+Strategies grid. HoRIM identifies and handles horizontal relations (e.g., conflicting and similar strategies) between strategies in different branches, but we have yet to adequately inspect the impact of HoRIM on identifying correct horizontal relations and improving grids. This lack of clarity hampers the application of HoRIM to industrial business strategy models. Herein, we evaluate the impact of HoRIM on the review process and the improvement process of GQM+Strategies grids using two experiments. The review experiment confirms that HoRIM identifies about 1.5 more horizontal relations than an ad hoc review. The modification experiment where four researchers evaluated the validity of improved grids by the ranking method suggests that HoRIM effectively modifies GQM+Strategies grids.

Keywords GQM+Strategies · Business strategy model · Horizontal Relation Identification Method

Y. Aoki (✉) · H. Washizaki · C. Shimura · Y. Senzaki · Y. Fukazawa
Information and Computer Science, Waseda University, Tokyo, Japan
e-mail: yoheiaoki1207@gmail.com

H. Washizaki
e-mail: washizaki@waseda.jp

C. Shimura
e-mail: chimaki.wsd@moegi.waseda.jp

Y. Senzaki
e-mail: yuitiro.senzaki@ruri.waseda.jp

Y. Fukazawa
e-mail: fukazawa@waseda.jp

© Springer International Publishing AG 2018
R. Lee (ed.), *Computer and Information Science*, Studies in Computational
Intelligence 719, DOI 10.1007/978-3-319-60170-0_4

43

1 Introduction

Aligning organizational goals and strategies is important in the Business Process Management (BPM) community [1]. Balanced Scorecard [2], Enterprise Resource Planning (ERP) Systems [3], and GQM+Strategies[®][1] [4, 5] support such an alignment.

GQM+Strategies provides a hierarchical structure called a GQM+Strategies grid based on the organizational structure. A GQM+Strategies grid is iteratively generated by decomposing the initial goal into strategies supporting goal achievement. The grid coordinates goals and strategies across different levels. However, GQM +Strategies grids may contain horizontal relations (e.g., conflicting strategies) between strategies in different branches. However, the GQM+Strategies framework does not provide a method to identify and handle horizontal relations.

Previously, we proposed the Horizontal Relation Identification Method (HoRIM) to iteratively improve a GQM+Strategies grid [6]. To identify horizontal relations, HoRIM detects differences between the initial GQM+Strategies grid and a model obtained by applying Interpretive Structural Modeling (ISM) [7] to the initial grid. Then, HoRIM provides a framework to modify the horizontal relations in a GQM+Strategies grid.

The effectiveness of HoRIM, particularly the improvement process, has yet to be thoroughly assessed, limiting the application of HoRIM to industrial business strategy models as well as the extension of HoRIM. Herein, we study the impact of HoRIM on the review process and the improvement process compared to the ad hoc method. Our experiments address the following research questions (RQs):

- RQ1: Does HoRIM more effectively identify horizontal relations in GQM +Strategies grids?
- RQ2: Does HoRIM improve the quality of GQM+Strategies grids?

To answer the above research question, this paper experimentally evaluates HoRIM using review and modification experiments. In the review experiment, subjects identified the horizontal relations from the GQM+Strategies grid with HoRIM or an ad hoc review. In the modification experiment, subjects suggested several alternatives to modify the horizontal relations and improve the GQM +Strategies grid. Then, evaluators who research GQM+Strategies ranked the improved grid.

The contribution of this paper is that we demonstrate the effectiveness of our method experimentally. The results lead to two main findings. First, HoRIM is effective in both the review process and the improvement process. Second, the evaluators cannot appropriately evaluate the strategy models when the comprehension of the background of the model is different between the evaluators and the proposer of it.

---

[1]GQM+Strategies[®] is a registered trademark (No. 302008021763) at the German Patent and the Trade Mark Office (international registration number IR992843).

2 Background

2.1 GQM+Strategies

The GQM+Strategies method is an extension of the GQM approach, which is used to create and establish measurement programs. The GQM+Strategies method also provides a hierarchical structure called a GQM+Strategies grid to align organizational goals and strategies at different levels.

A GQM+Strategies grid consists of GQM graphs [8] and GQM+Strategies elements (Fig. 1). The GQM graph monitors all goals at various levels of an organization to evaluate the achievement of each goal. The graph involves three concepts: goals, questions, and metrics. GQM+Strategies elements align goals and strategies throughout an organizational hierarchy. These elements specify organizational goals, strategies, rationales, and their relationships.

The GQM+Strategies framework creates a grid by repeatedly defining lower-level goals and strategies based on the initial set of goals and strategies. That is, the GQM+Strategies grid is specified from the initial goal, which is repeatedly decomposed to create a concrete goal. Generating a GQM+Strategies grid is three-step process: (1) define the initial goal; (2) specify the strategies to achieve the goal and the rationales to explain how the strategies will realize the goal; and (3) define the goals of the lower-level units and return to step 2.

2.2 Horizontal Relations

Our efforts focus on the strategies in GQM+Strategies grids. We define a vertical relationship as a parent–child relation between strategies. Although GQM+Strategies grids frequently have horizontal relations between strategies in different branches (Fig. 2), the GQM+Strategies method does not support horizontal relations.

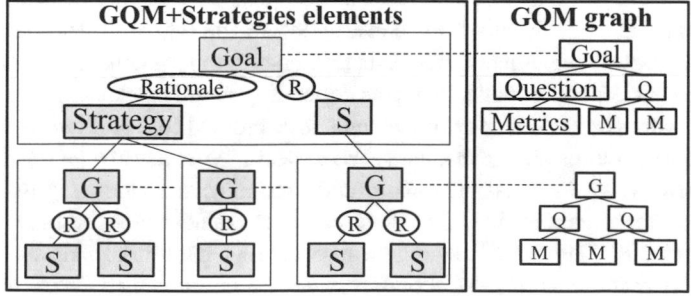

Fig. 1 GQM+Strategies grid

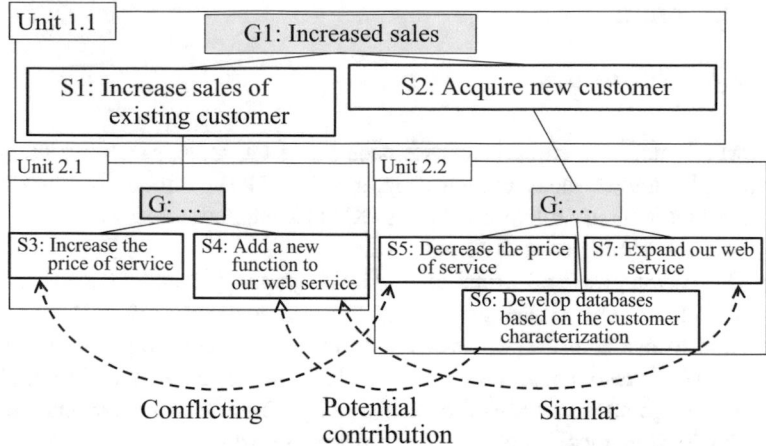

Fig. 2 Example of a GQM+Strategies grid with horizontal relations

Horizontal relations can be classified into three categories:

- Conflicting strategies, which contradict or negatively influence each other. Conflicting strategies must be identified and resolved in order for an organization to run smoothly and effectively.
- Potential contributions, where one strategy contributes to a strategy in another branch. An organization should identify potential contributions to other strategies to improve the quality of its products.
- Similar strategies, which are executed by the same approach or have the same target. To improve efficiency, similar strategies should be identified and merged.

2.3 Related Work

John N. Warfield developed the ISM approach, which generates a hierarchical structure to analyze relationships between elements in complex systems [7]. The hierarchical structure visualizes the construction of the whole system based on the dependence between elements. Elements influencing other elements are placed in a lower layer. On the other hand, elements depending on other elements are placed in a higher layer. The dependency of elements is expressed in a relation matrix where the rows and columns are the elements. We expect ISM to assist a GQM+Strategies grid reviewer (e.g., business analyst) in understanding and analyzing models.

Conflict management is field of research to handle and resolve conflicts. T. Ruble and K. Thomas [9] identified five conflict handling modes: competing, avoiding, accommodating, collaborating, and compromising. These modes are classified based on whether the opposing persons are assertive or cooperative. We expect the theory of conflict management to be applicable to conflicting strategies, similar strategies, and potential contributions.

Several researchers have struggled to improve business process models. M. E. Khalaj et al. suggested a semantic framework to model business process based on software architectural concepts, which significantly reduces the misunderstanding of complexities [10]. W. Khlif supposed that combining the semantic aspect with the structural aspect further reduces the control flow complexity of a business process modeled in the Business Process Modeling Notation [11].

The GQM+Strategies method has been expanded. T. Kobori et al. suggested the Context-Assumption-Matrix (CAM), which refines the GQM+Strategies model by extracting rationales based on analyzing the relationships between stakeholders [12, 13]. C. Shimura defined modeling rules for GQM+Strategies with a metamodel and design principles that consist of relationship constraints between GQM+Strategies elements [14]. This method helps identify and improve potential problems and strategic risks. To develop a strategy that considers the requirements of both the user and the business, C. Uchida et al. described the GO-MUC method (Goal-oriented Measurement for Usability and Conflict) [15].

3   Horizontal Relation Identification Method (HoRIM)

Previously, we proposed HoRIM [6] to identify and handle horizontal relations of strategies in GQM+Strategies grids. Figure 3 overviews HoRIM. After constructing a GQM+Strategies grid, HoRIM is used as a review. HoRIM consists of the following steps: reconstruction, analysis, and modification. In the reconstruction phase, the hierarchical structure involving horizontal relations is generated by ISM. In the analysis phase, HoRIM detects the differences between an initial GQM +Strategies grid and the hierarchical structure. Then, the identified horizontal relations are classified into three categories: conflicting strategies, similar strategies, and potential contributions. In the modification step, several alternatives to deal with horizontal relations based on the five-modification approach are suggested and examined. Finally, the GQM+ Strategies grid is improved according to the alternatives. This process is iteratively executed to address horizontal relations and improve the GQM+ Strategies grid.

3.1   Reconstruct

In this step, the hierarchical structure consisting of the strategies is generated by ISM. ISM uses a relation matrix to determine the dependency between any two elements. An analyst creates relation matrix A = {a_{ij} | i, j = 1, 2, ..., n} to express all direct binary relationships. "n" means the number of the strategies, which are the rows and columns of the relation matrix. If the column element depends on the row

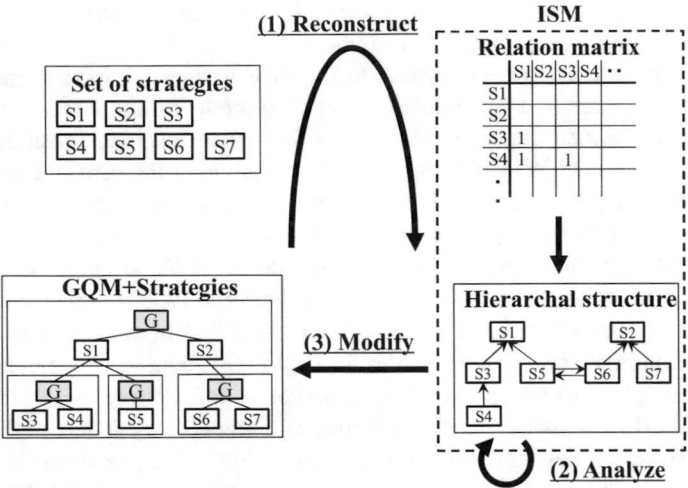

Fig. 3 Overview of HoRIM

element, a value of 1 is inputted. Otherwise, 0 is inputted. The analyst specifies all relations (involving horizontal relations) between strategies into the relation matrix. Then, the hierarchical structure is automatically generated by the algorithm as ISM.

3.2 Analyze

In this step, horizontal relations are identified by the hierarchical structure. ISM generates the hierarchical structure from all dependencies between the strategies, whereas the GQM+Strategies grid is constructed based on top-down approach. Therefore, the hierarchical structure of ISM expresses the relations that the GQM+Strategies grid cannot specify. To identify horizontal relations, HoRIM detects the difference between the initial GQM+Strategies grid and the hierarchical structure by ISM.

3.3 Modify

Finally, the GQM+Strategies grids are modified from the viewpoint of horizontal relations. HoRIM employs the five-modification approach: detail, select, integrate, breakthrough, and relate.

Detail means that applying strategies concretely prevents the overlap of strategic objects. This is particularly effective when strategies are described abstractly. Select means to compare two or more strategies before choosing one. Select is effective when the strategic priorities differ significantly. Integrate means to combine two strategies into one unified strategy. An integrated strategy often becomes an abstract version of the original ones. Breakthrough means to create new strategies to resolve conflicting strategies. Techniques such as a conflict resolution diagram [16] can be utilized to discover a new strategy. Relate means that the relation between the strategies with horizontal relations is added.

When several modification alternatives are considered, they are examined based on the following viewpoints: certainty of solving the problem, contribution toward goals, potential negative effect, and obstacles for execution. From the viewpoint of certainty, the analyst answers how the modification alternative resolves the problem (e.g., non-efficiency by conflicting strategies or dispersion of business process by similar strategies). From the viewpoint of the contribution to goals, the analyst should confirm that the alternative does not interrupt the original goals in GQM +Strategies grid. A potential negative effect means the alternative adversely influences other strategies. Alternatives with negative effects may induce other horizontal relations. An execution obstacle indicates a difficulty or complexity of reorganization or the new strategy that the alternative specifies.

4 Evaluation

4.1 Experiment Planning

We compared the effectiveness of identifying and modifying horizontal relations by HoRIM and an ad hoc review, which is subjectively executed. During our evaluation, we investigated the research questions described in Sect. 1. To answer the research questions, we conducted two experiments on GQM+Strategies grids. One reviewed the grid, while the other modified it. Table 1 overviews our experiments.

The review experiment involved six university students majoring in computer sciences. All students were familiar with how to model GQM+Strategies grids. The students were divided into two groups of three students (Groups A and B). Each subject completed two exercises, where he or she identified the horizontal relations from the real GQM+Strategies grid with HoRIM or ad hoc review. To reduce the learning effects, Group A completed exercise 1 by HoRIM, while Group B executed exercise 1 by an ad hoc review. In exercise 2, the methods were reversed for each group. Both GQM+Strategies grids included 3 level layers and had 23 strategies. We measured the number of identified horizontal relations in this experiment.

The modification experiment involved university students majoring in computer sciences. The 12 subjects were divided randomly into two groups of six (Groups A and B). Group A performed exercise 3 by HoRIM, while Group B executed exercise 3 by an ad hoc method. In exercise 4, the methods were reversed.

The materials of the grid already specified three horizontal relations: conflicting strategies, similar strategies, and potential contribution. Firstly, the subjects

Table 1 Overview of our experiments

	Purpose	Task	Target	Time	Method	
					Group A	Group B
Ex. 1	To evaluate the effectiveness of identifying horizontal relations	Identify horizontal relations from the GQM+Strategies grid	6 students	Individual	HoRIM	Ad hoc
Ex. 2	To evaluate the effectiveness of identifying horizontal relations	Identify horizontal relations from the GQM+Strategies grid	6 students	Individual	Ad hoc	HoRIM
Ex. 3	To evaluate the effectiveness of modifying horizontal relations	Improve the GQM +Strategies grid	12 students	50 min	HoRIM	Ad hoc
Ex. 4	To evaluate the effectiveness of modifying horizontal relations	Improve the GQM +Strategies grid	12 students	50 min	Ad hoc	HoRIM

suggested all modification alternatives that they could envision. Then, they modified and improved the GQM+Strategies grid to deal with horizontal relations. The grids in this experiment differed from the ones in the review experiments. The material grids were constructed based on two industrial cases. The grids were simple as they included two level layers and seven strategies. We measured the number of modification alternatives.

Four researchers, who studied the GQM+Strategies framework or the business models, evaluated the validity of the modified grids. They ranked the modified grids because it is difficult to estimate the absolute validity. The evaluators could not give the same rank to different objects.

4.2 Results

Figure 4 shows the results of the review experiment. The precision and recall were calculated from the identified horizontal relations and the correctly defined horizontal relations. Table 2 shows the results of the modification experiment of the GQM+Strategies grids and the evaluation of the modified grids. Num. stands for the number of modification alternatives and the evaluator rows show the rank of the modified grids. Attr. stands for the integration value of the evaluation by Thurstone's method [17, 18], which converts an ordinal scale into an interval scale, assuming that the quality of the samples follows a normal distribution. A high figure means a high-quality modified grid, while a low figure means the

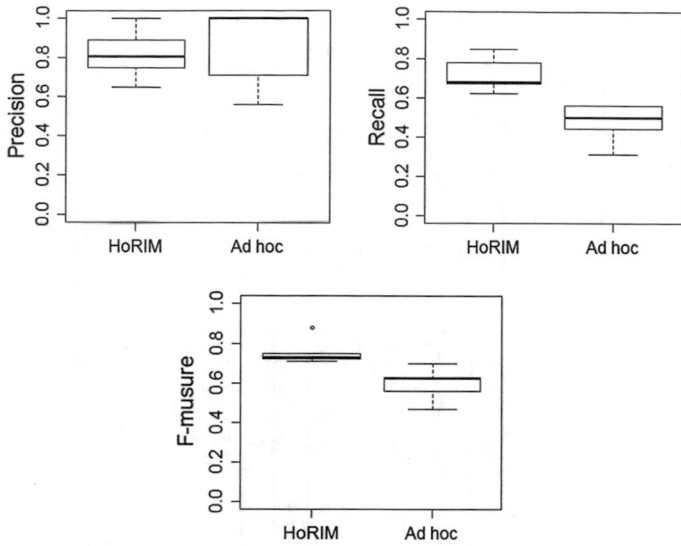

Fig. 4 Boxplots of the experiment reviewing GQM+Strategies grid

low-quality. Figure 5 shows the boxplot diagram of the number of suggested alternatives and the result of Thurstone's method. Table 3 shows Kendall's coefficient of concordance [19] and the results of Mann–Whitney U-test.

4.3 Discussion

4.3.1 RQ1

The average recall of HoRIM is about 1.48 times that of the ad hoc review (Table 2), confirming that HoRIM is more effective. Subjects using HoRIM suggested relations involving three or more strategies, whereas the ad hoc review identified relations between two strategies. These findings indicate that HoRIM helps understand more complex strategies, confirming that it assists in analyzing complex GQM+Strategies grids.

The precision of HoRIM is lower than that of the ad hoc review for the cosmetic company in exercise 1. In addition, the group using the ad hoc review in exercise 1 made more mistakes in exercise 2 using HoRIM. These results imply that not all horizontal relations suggested by HoRIM are correct. However, the significant difference is not observed (Table 3).

Table 2 Results of the experiment modifying GQM+Strategies grids

		Subjects											
		1	2	3	4	5	6	7	8	9	10	11	12
Ex. 3	Method	HoRIM						Ad hoc					
	Num.	4	10	8	6	6	7	5	6	3	4	2	4
	Evaluator 1	10	5	4	12	1	3	2	6	8	9	11	7
	Evaluator 2	12	1	5	6	2	7	4	3	8	10	11	9
	Evaluator 3	2	3	5	4	8	1	9	7	10	11	12	6
	Evaluator 4	1	4	9	3	7	2	10	8	6	11	12	5
	Attr.	0.65	0.98	0.68	0.59	0.88	1.00	0.65	0.67	0.47	0.23	0.00	0.59
Ex. 4	Method	Ad hoc						HoRIM					
	Num.	2	4	5	4	5	4	7	6	7	1	3	3
	Evaluator 1	9	10	2	8	7	12	1	3	5	6	11	4
	Evaluator 2	3	2	11	6	10	5	4	7	8	9	12	1
	Evaluator 3	12	4	11	1	3	10	5	2	7	8	6	9
	Evaluator 4	2	3	5	1	4	7	6	10	12	11	9	8
	Attr.	0.46	0.76	0.38	1.00	0.57	0.15	0.92	0.67	0.23	0.22	0.00	0.72

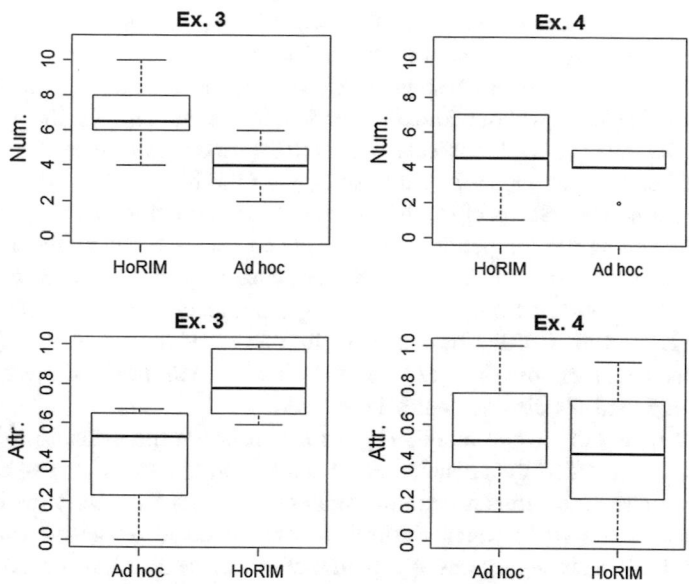

Fig. 5 Boxplots of the experiment modifying GQM+Strategies grids

Table 3 Results of the test

		Kendal	Wilcox
Ex. 1 & 2	Precision	–	0.423
	Recall	–	0.004
	F-measure	–	0.004
Ex. 3	Num.	–	0.030
	Rank	0.035	0.030
Ex. 4	Num.	–	0.762
	Rank	0.396	0.699

4.3.2 RQ2

In exercise 3, we confirmed that HoRIM is effective as the subjects with HoRIM suggested more modification alternatives and higher-quality modified grid. On the other hand, in exercise 4, we found no significant difference between the number of alternatives. In addition, the rank of the modified grid indicates no concordance between the evaluators (Tables 2 and 3). It is considered that the subjects using ad hoc review in exercise 4 had already learned HoRIM in exercise 3. In fact, we confirmed concept words of HoRIM (e.g., detail or select) in the answers of the group that did not use HoRIM in exercise 4. Individual differences seem to be low as significant differences are not found in exercise 4 despite finding significant differences in exercise 3. The effectiveness of HoRIM does not seem to depend on the complexity or size of the GQM+Strategies grid as we designed the grids to have the same size, complexity, and types of horizontal relations. In conclusion, HoRIM appears to effectively modify GQM+Strategies grids.

The subjects using HoRIM suggested modification alternatives from various viewpoints. Subjects without HoRIM tended to delete one of the conflicting strategies (i.e., select), while subjects with HoRIM tried to coordinate the strategies (i.e., detail or breakthrough). In exercise 3, the GQM+Strategies grid contained conflicting strategies; S1 specifies the increment of competition participants, while S2 specifies setting the competition theme. This conflict is based on the assumption that only people interested in the theme participate the competition. Subject 11 without HoRIM deleted S2 upon considering the impact to its goal. On the other hand, subject 5 with HoRIM modified S2 into a detailed strategy specifying that participants could choose between several themes. Similarly, subject 11 with HoRIM suggested detailed strategies in exercise 4.

The evaluation of several modified grids depended on the evaluators. In particular, the grids modified by a detailed approach tended to receive dispersed evaluations. In exercise 3, the grids had similar strategies. S3 and S4 specify the increment of feedback to the participants. Subject 1, who received dispersed evaluations, suggested detailed strategies; one specified that proper feedback is increased, while another specified that participants receive feedback quickly. One evaluator judged that the two strategies represented two viewpoints (quality and delivery), whereas another evaluator felt that they were the same as both improved feedback.

How the modified grids are assessed is one cause of the difference in the evaluation. One evaluator used the potential of the modified grid to determine the validity. Another evaluator assessed the grids based on his or her perception of what the correct grid should be. The latter approach will result in poor marks if the grid is inconsistent with the evaluator's expectations regardless of the grid quality. In our experiment, it is presumed that we introduced the background and premise of the GQM+Strategies grid's domain to the evaluators and subjects relatively well in exercise 3, but not in exercise 4, leading to a misunderstanding in exercise 4.

4.4 Findings and Their Usage

Our experiments revealed the following three findings:

- The review experiment demonstrated that a structural analysis method such as ISM can effectively identify misalignments in the strategy models. Therefore, the business analyst should utilize HoRIM or a structural analysis method when analyzing complex and large strategy models.
- Exercise 3 confirmed that the concept of the modification for horizontal relations leads to more modification alternatives and the proper improvement of the strategy models. Therefore, the business analyst should consider the modification approach and the evaluation viewpoints in HoRIM when improving the strategy models.

- Exercises 3 and 4 indicated that the evaluation of strategy models depends on the background and promise of the domain. Therefore, the researchers should devise a method to reconcile the background and promises of the amenders and the evaluators when validating strategy models.

4.5 Threats to Validity

One threat to internal validity is the difference between the abilities and experiences of the subjects. However, this bias was removed by dividing the subjects into two random groups. For exercises 1 and 3, Group A employed HoRIM, while Group B used an ad hoc review. The employed methods were reversed in exercises 2 and 4. Exercises 1, 2, and 3 demonstrate that HoRIM is more effective than an ad hoc review. However, the small sample size cannot confirm the precision or effectiveness of HoRIM. In the future, an experiment involving a larger sample size is necessary.

There are two threats to external validity. First, the subjects were students with limited knowledge of the strategies in the GQM+Strategies grids. Second, only two GQM+Strategies grids were examined in each experiment. The small number of strategies may decrease HoRIM's superiority because simple GQM+Strategies grids are easily analyzed.

5 Conclusion and Future Work

GQM+Strategies grids frequently contain relations between the strategies in different branches. Such relations are defined as horizontal relations. To handle horizontal relations, we proposed the Horizontal Relation Identification Method (HoRIM) to detect the difference between the initial GQM+Strategies grid and a model by ISM. HoRIM provides a framework to improve grids. Our experiment demonstrates that HoRIM improves the effectiveness of not only identifying horizontal relations but also modifying GQM+Strategies grids.

In the future, we plan to replicate our experiment modifying the GQM +Strategies grids to consider the learning effects. Additionally, we plan to expand HoRIM so that it can distinguish the types of relations (e.g., positive, negative, and overlap), which should improve the analysis of hierarchical structures.

References

1. Burlton, R.: Delivering business strategy through process management. In: Handbook on Business Process Management, vol. 2, pp. 5–37. Springer, Berlin, Heidelberg (2010)
2. Kaplan, R.S., et al.: The balanced scorecard: measures that drive performance. Harv. Bus. Rev. **83**(7) (2005)
3. Mulazzani, F., Russo, B., Succi, G.: ERP systems development: enhancing organization's strategic control through monitoring agents. In: IEEE/ACIS ICIS (2009)
4. Basili, V.R., et al.: Linking software development and business strategy through measurement (2013). arXiv:1311.6224
5. Basili, V.R., et al.: Bridging the gap between business strategy and software development. In: ICIS (2007)
6. Aoki, Y., Washizaki, H., et al.: Identifying misalignment of goal and strategies across organizational units by interpretive structural modeling. In: HICSS (2016)
7. Warfield, J.N.: Intent structures. IEEE Trans. Syst Man Cybern. **2**, 133–140 (1973)
8. Van Solingen, R., et al.: Goal question metric (GQM) approach. In: Encyclopedia of Software Engineering (2002)
9. Ruble, T.L., Thomas, K.W.: Support for a two-dimensional model of conflict behavior. Organ. Behav. Hum. Perform. **16**(1), 143–155 (1976)
10. Khalaj, M.E., et al.: A semantic framework for business process modeling based on architecture styles. In: IEEE/ACIS ICIS (2012)
11. Khlif, W., Ben-Abdallah, H.: Integrating semantics and structural information for BPMN model refactoring. In: IEEE/ACIS ICIS (2015)
12. Kobori, T., Washizaki, H., et al.: Efficient identification of rationales by stakeholder relationship analysis to refine and maintain GQM+Strategies models. In: APRES (2014)
13. Kobori, T., Washizaki, H., et al.: Identifying rationales of strategies by stakeholder relationship analysis to refine and maintain GQM+Strategies models. In: PROFES (2014)
14. Shimura, C., Washizaki, H., et al.: Identifying potential problems and risks in GQM +Strategies models using metamodel and design principles. In: HICSS (2017)
15. Uchida, C., Washizaki, H., et al.: GO-MUC: a strategy design method considering requirements of user and business by goal-oriented measurement. In: CHASE (2016)
16. William Dettmer, H.: The conflict resolution diagram: creating win-win solutions. J. Qual. Particip. **27**(2) (2004)
17. Thurstone, L.L.: Attitudes can be measured. Am. J. Sociol. **33**(4), 529–554 (1928)
18. Leon Harter, H.: Expected values of normal order statistics. Biometrika **48**(1/2) (1961)
19. Kendall, M.G., Babington Smith, B.: The problem of m rankings. Ann. Math. Stat. **10**(3) (1939)

Combining Lexicon-Based and Learning-Based Methods for Sentiment Analysis for Product Reviews in Vietnamese Language

Son Trinh, Luu Nguyen and Minh Vo

Abstract Social media websites are a major hub for users to express their opinions online. Businesses spend an enormous amount of time and money to understand their customer opinions about their products and services. Sentiment analysis which is also called opinion mining, involves in building a system to collect and examine opinions about the product made in blog posts, comments, or reviews. In this paper, we propose a framework for sentiment analysis based on combining lexicon-based and learning-based methods for product review sentiment analysis in Vietnamese language. Text analytics, Linguistic analysis and Vietnamese emotional dictionary were built, proposing features which adapted with the language was proposed. The experimental show that our system has very well performance when combine advantage of lexicon-based and learning based and can be applied in online systems for sentiment analysis product reviews.

Keywords Lexicon-based · Learning-based · Sentiment analysis · Vietnamese · Text analytics · Linguistic analysis · Vietnamese emotional dictionary · Proposing features · Product review

1 Introduction

Social media websites are a major hub for users to express their opinions online. On these social media sites, users post comments and opinions on various topics. Hence these sites become rich sources of information to mine for opinions and

S. Trinh (✉) · L. Nguyen · M. Vo
University of Information Technology, Ho Chi Minh City, Vietnam
e-mail: sontq@uit.edu.vn

L. Nguyen
e-mail: luut.ng@gmail.com

M. Vo
e-mail: voleminh10t2@gmail.com

© Springer International Publishing AG 2018   57
R. Lee (ed.), *Computer and Information Science*, Studies in Computational
Intelligence 719, DOI 10.1007/978-3-319-60170-0_5

analyze user behavior and provide in-sights for user behavior, product feedback, user intentions, lead generation.

Sentiment analysis is a type of natural language processing for tracking the mood of the public about a particular product or topic. Sentiment analysis, which is also called opinion mining, involves in building a system to collect and examine opinions about the product made in blog posts, comments, or reviews. Sentiment analysis can be useful in several ways. For example, in marketing it helps injudging the success of an ad campaign or new product launch, determine which versions of a product or service are popular and even identify which demo graphics like or dislike particular features. Thus sentiment analysis has become a hot research area. Sentiment Analysis is used to determine sentiments, emotions and attitudes of the user. Businesses spend an enormous amount of time and money to understand their customer opinions about their products and services.

Lexicon-based approaches to sentiment analysis differ from the more common machine-learning based approaches in that the former rely solely on previously generated lexical resources that store polarity information for lexical items, which are then identified in the texts, assigned a polarity tag, and finally weighed, to come up with an overall score for the text. In [1], we used this method for sentiment analysis of Facebook comments in Vietnamese and had the good performance in the experimental. However, those types of online data have several flaws that potentially hinder the process of sentiment analysis. The first flaw is that since people can freely post their own content, the quality of their opinions cannot be guaranteed.

This paper continues previous research by Son Trinh et al. (2016), which implemented sentiment analysis of Facebook comments in Vietnamese language, when we applied in business domain to sentiment analysis for product, we recognize that a sentiment analysis system will be the best if it must be included linguistic analysis module which can be deep support in text analysis, especially must be recognize phrases exactly in sentences, where are important phrases and their role in sentence based on grammar; the secondly, features in training is very important and affect to result in domain and combining lexicon-based and learning-based methods for product reviews sentiment analysis in Vietnamese language is a good way to improve the accuracy of sentiment classification because of advantage of their. The experimental show that our system has very well performance and can be applied in online systems for sentiment analysis product reviews.

The rest of the paper is organized as follows. In Sect. 2, we present related work on sentiment analysis and characteristics of Vietnamese language in Sect. 3 and then propose a framework for sentiment analysis based on combining lexicon-based and learning-based methods for product review sentiment analysis in Sect. 4. In Sect. 5, we present experimental evaluate the final result obtained from the test data. In Sect. 6, we present our conclusions and outline future work.

2 Related Work

Sentiment Analysis on raw text is a well known problem. The Liu [2] book covers the entire field of sentiment analysis. Sentiment analysis can be done using machine learning, lexicon-based approach or combined.

The machine learning approach applicable to sentiment analysis mostly belongs to supervised classification in general and text classification techniques in particular [1]. However, their obvious disadvantage in terms of functionality is their limited applicability to subject domains other than the one they were designed for. In a machine learning based classification, two sets of documents are required: training and a test set. A training set is used by an automatic classifier to learn the differentiating characteristics of documents, and a test set is used to validate the performance of the automatic classifier. A number of machine learning techniques have been adopted to classify the reviews. Machine learning techniques as Naive Bayes, maximum entropy and support vector machines (SVM) [1]. Although interesting research has been done aimed at extending domain applicability [3], such efforts have shown limited success. An important variable for these approaches is the amount of labeled text available for training the classifier, although they perform well in terms of recall even with relatively small training sets [4]. On the other hand, a growing number of initiatives in the area have explored the possibilities of employing unsupervised lexicon-based approaches.

The semantic orientation approach to sentiment analysis is unsupervised learning because it does not require prior training in order to mine the data. Instead, it measures how far a word is inclined towards positive and negative. Much of the research in unsupervised sentiment classification makes use of lexical resources [5]. The lexicon based approach is based on the assumption that the contextual sentiment orientation is the sum of the sentiment orientation of each word or phrase. Turney [6] identifies sentiments based on the semantic orientation of reviews. Taboada [7], Melville [8], Ding [9] use lexicon based approach to extract sentiments.

These rely on dictionaries where lexical items have been assigned either polarity or a valence, which has been extracted either automatically from other dictionaries, or, more uncommonly, manually. The works by Hatzivassiloglou and McKewon [10] and Turney [6] are perhaps classical examples of such an approach. The most salient work in this category is Taboada [7], whose dictionaries were created manually and use an adaptation of Polanyi and Zaenen's [11] concept of Contextual Valence Shifters to produce a system for measuring the semantic orientation of texts, which they call SOCAL.

Combining both methods (machine learning and lexicon-based techniques) has been explored by Kennedy and Inkpen [12], who also employed contextual valence shifters, although they limited their study to one particular subject domain. The degree of success of knowledge based approaches varies depending on a number of variables, of which the most relevant is no doubt the quality and coverage of the lexical resources employed, since the actual algorithms employed. However by using combine, we're get more advantages from two methods.

Currently, the research of sentiment problem in Vietnamese language has some results. In particular, Nguyen [13] show up the problem in building a sentiment dictionary is difficult and time consuming. An approach to mining public opinions from Vietnamese text using a domain specific sentiment dictionary in order to improve the accuracy. The sentiment dictionary is built incrementally using statistical methods for a specific domain [14].

In another way, in [15, 16] authors have explored different methods of improving the accuracy of sentiment classification. The sentiment orientation of a document can be positive (+), negative (−), or neutral (0). Dictionary has many verbs, adverbs, phrases and idioms. The author based on the combination of term-counting method and enhanced contextual valence shifters method has improved the accuracy of sentiment classification. The combined method has accuracy 68.9% on the testing dataset, and 69.2% on the training dataset. All of these methods are implemented to classify the reviews based on our new dictionary and the internet movie data set. The result is still low performance and can not applied in real systems.

We used lexicon based method for sentiment analysis of Facebook comments in Vietnamese language and had good performance in the experimental [1]. However, those types of online data have several flaws that potentially hinder the process of sentiment analysis. The first flaw is that since people can freely post their own content, the quality of their opinions cannot be guaranteed.

When we applied in bussiness domain for sentiment analysis product reviews, we recognize that a sentiment analysis system will be the best if it must be included linguistic analysis module which can be deep support in text analysis, especially must be recognize phrases exactly in sentences, where are important phrases and their role in sentence based on grammar; the secondly, features in training is very important and affect to result in domain and combining lexicon-based and learning-based methods is a good way to improve the accuracy of sentiment classification because of advantage from methods.

3   Characteristics of Vietnamese Language

Vietnamese is a monosyllable, tonal language. Each word unit is pronounced as a syllable and its meaning depends on the tone. There are about 6,596 phonetically distinguishable syllables which comprise of legal combinations between basic syllables (i.e. syllables without tone) and five tones. Table 1 illustrates the diacritics used for representing tones, including: level tone (denoted by "none"), high-rising tone (/), low-falling tone (\), dipping-rising tone (?), high-rising glottalized tone (~), and low glottalized tone (.). Although word, a group of one to several syllables is the smallest syntactically meaningful unit, syllable is the basic pronunciation unit in Vietnamese speech.

Table 1 Diacritics in Vietnamese [18]

Diacritic	None	/	\	?	~	.
Example	xa	xá	xà	xả	xã	xạ
Meaning	far	bow	snake	release	village	musk

Table 2 Classification table of Vietnamese vowel—Monophthongs [18]

	Front	Central	Back
Close	I	i	U
Close-mid	E	ə:	O
Open-mid	ɛ	ə	ɔ
Open		a a:	

Table 3 Classification table of Vietnamese vowel—Diphthongs and triphthongs [18]

/ə/ Diphthongs	/j/ Diphthongs/ Triphthongs	/w/ Diphthongs/ Triphthongs
/iə/	/ɔ:j/	/iw/
/iə/	/ɔj/	/ew/
/uə/	/a:j/	/ɛw/
	/aj/	/ɔ:w/
	/ij/	/ɔw/
	/ij/	/ɔw/
	/uj/	/a:w/
	/oj/	/aw/
	/ɔj/	/iw/
	/iɔj/	/iɔw/
	/uɔj/	/iɔw/

Tables 2 and 3 shows Vietnamese vowels while Table 4 show consonants. Phonemes are combined to form a syllable, and several syllables are combined to form a word phrase which is different from a phrase in Vietnamese.

Table 4 Classification table of Vietnamese consonant [18]

		Bilabial	Labiodental	Dental/Alveolar	Post-alveolar	Palatal	Velar	Glottal
Nasal		M		N		ɲ	ŋ	
Plosive and Affricate	Unaspirate d	P		T	tʂ	C	k	(ʔ)
	Aspirated			tʰ				
	Glottalized	ɓ		ɗ				
Fricative			f(vʲ)	s	ʂʐ		x ɣ	H
Approximant				l		j	w	

The Vietnamese syllable structure is: $(C_1)(w)V(C_2) + T$ where $C_1 =$ initial consonant onset, $w =$ bilabial glide /w/, $V =$ vowel nucleus, $C_2 =$ final consonant coda, $T =$ tone.

Vietnamese is an analytic language, meaning it has no inflection of any kind. The sentence structure in Vietnamese which expresses the idea that an action or event is happened routinely or habitually. Structure: Subject (S) + Verb (V) + Object (O). For example: *I like this product (Tôi thích sản phẩm đó)*.

Negative Form

To formulate a negative sentence, "không" or "không phải" is placed before the main verb of the sentence. Structure: S + không/không phải + V + O. For Example: *I don't like this product (Tôi không thích sản phẩm đó)*.

Interrogative Form

To form an interrogative sentence, "không" is placed at the end of the sentence. To make the sentence sound more natural, "có" is placed before the main verb. Structure: S + có + V + O + không?. For example: *Do you like this product? (Bạn có thích sản phẩm đó không?)*.

By the way, Vietnamese has rules to rules for sentence structures to know how many clauses in sentences, which is the main clause and what is the meaning. For example: If A then B or If A, B (A thì B). In this structure, we have two clauses (A, B) in sentence and B is the result clause.

4 Sentiment Analysis for Product Reviews

Comparing with previous researches related to our topic, our propose method has different points, that are features which were selected adaptation in Vietnamese language, built a Vietnamese emotional language with more words consistent with the Vietnamese grammar based on spelling that people are using. We also built a linguistic analysis which can be deep support in text analysis and combining lexicon-based and learning-based be proposed.

4.1 Data Collection

We collect all comments or reviews automatically from the technology sites

4.2 Sentiment Sentences Extraction and POS Tagging

It is suggested by Pang and Lee [17] that all objective content should be removed for sentiment analysis. Instead of removing objective content, in our study, all

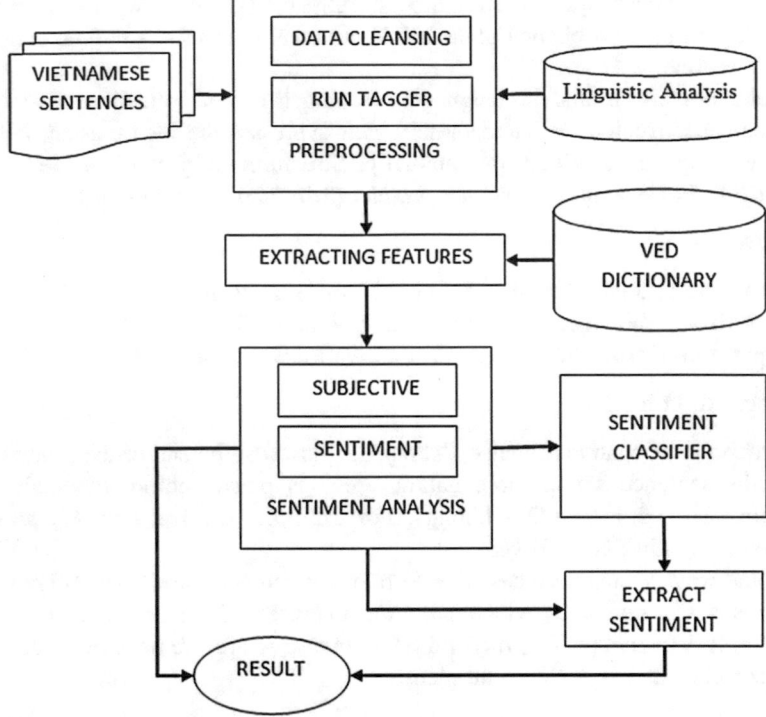

Fig. 1 Our proposed system

subjective content was extracted for future analysis. The subjective content consists of all sentiment sentences. A sentiment sentence is the one that contains, at least, one positive or negative word. All of the sentences were firstly tokenized into separated words. Every word of a sentence has its syntactic role that defines how the word is used (Fig. 1).

The syntactic roles are also known as the parts of speech. There are 8 parts of speech in Vietnamese: verb, noun, pronoun, adjective, adverb, preposition, conjunction, and interjection. In this component, we used linguistic analysis module for identifying how many phrases in sentences, which is important phrase and their role in sentence based on Vietnamese grammar. Linguistic analysis module implemented using rules which created manually by linguistic experts for sentence types.

4.3   Building Emotional Dictionary

The calculation of sentiment based on dictionary begins with two assumptions: that individual words have what is referred to as prior polarity, that is, a semantic orientation that is independent of context and that said semantic orientation can be expressed as a numerical value. Several lexicon based approaches have adopted these assumptions.

Adjectives: Much of researches in sentiment focused on adjectives or adjective phrases as the primary source of subjective content in a document. In general, the semantic orientation of an entire document is the combined effect of the adjectives or relevant words found within, based upon a dictionary of word rankings (scores) [13].

Nouns, Verbs, and Adverbs: Although the sentences have comparable literal meanings, the plus-marked nouns, verbs, and adverbs indicate the positive orientation of the speaker towards the situation, whereas the minus-marked words have the opposite effect. It is the combination of these words in each of the sentences that conveys the semantic orientation for the entire sentence [5]. In order to make use of this additional information, we created separate noun, verb and adverb dictionaries, hand-ranked using the same +5 to −5 scale as our adjective dictionary. We created Vietnamese emotional dictionary (VED) which contains 5 sub-dictionaries by manually, including noun, verb, adjective, and adverb dictionary. Our dictionary is a part based on the English SO-CAL (Dictionaries for the Semantic Orientation CALculator) dictionary. We choose SO-CAL, because this dictionary is the best in overall for a lots of topic in experiments as shown [1].

We added some words to our dictionary to make consistent with the Vietnamese grammar and concise spelling that people are using on product reviews, and a part emotional words which appear in Vietnamese using in everyday life by people. The number of words in each dictionary of noun, verb, adjective and adverb are 4546 words, 4248 words, 5357 words, 3749 words respectively and each word is paired with an integer which describes the corresponding emotional value from the most negative (−5) to the most positive (+5). Notice that no word has emotional value at zero value (0).

In addition, we built dictionary for emotional icon which usually used in product reviews (Tables 5, 6, 7, 8, 9 and 10).

The intensification dictionary has 456 special words in Vietnamese language and each word also has a accompanied decimal to demonstrate the increase or decrease of its emotional value.

Example: If emotional value for word "nhếch nhác" (messy) is (−3) then word "khá nhếch nhác" (rather messy) has emotional value (−3) * (1 − 0.1) = (−2.7). On the same, if emotional value for word "xuất sắc" (excellent) is (5) then word "xuất sắc nhất" (the most excellent) has emotional value 5 * (1 + 1) = 10.

Table 5 Some emotional icon in product reviews

Emotional icon	Emotional value
:)~ Smile	1
:]~ Smile	2
=)~ Smile	4
:(~ Frown	−1
/~ Unsure	−1
o.O~ Confused −1	−1
>:-o~ Upset	−2

Table 6 Some words from dictionary of noun [1]

Noun	
Noun	Emotional value
hoàn hảo (perfection)	5
lộng lẫy (luxury)	4
chiến thắng (victory)	3
phước lành (blessing)	2
độc lập (liberty)	1
tội phạm (crime)	−1
điểm yếu (weakness)	−2

Table 7 Some words from dictionary of verb [1]

Verb	
Verb	Emotional value
tôn kính (respect)	4
hoan hỉ (delight)	4
thành công (succeed)	3
sáng tạo (create)	2
Tăng (increase)	1
vùi dập (ruin)	−1
xấu hổ (shame)	−2

Table 8 Some words from dictionary of adjective [1]

Adjective	
Adjective	Emotional value
tuyệt vời (perfect)	5
cao cấp (high-grade)	4
bổ ích (helpful)	3
chặt chẽ (close)	2
hợp lý (agreed)	1
cũ (old)	−1
đần độn (silly)	−2

Table 9 Some words from dictionary of adverb [1]

Adverb	
Adverb	Emotional value
thú vị (interestingly)	5
huy hoàng (splendidly)	4
giỏi (well)	3
Tươi (freshly)	2
sạch (clean)	1
kỳ quặc (weirdly)	−1
Thô (crudely)	−2

Table 10 Some words from intensification dictionary [1]

Intensification	Emotional value
Ít (Slenderly)	−1.5
chút ít (Slightly)	−0.9
hơi (a little)	−0.5
khá (rather)	−0.2
chắc (surely)	0.2
siêu (super)	0.4
hoàn toàn (completely)	0.5

Table 11 Some sentence structures in Vietnamese

ID	Sentence structure
1	Nếu A, B (If A, B)
2	Nếu A hoặc B (If A or B)
3	Bởi vì A nên B (Because A, B)
4	Không chỉ A, mà B (Not only A, but B)
5	Nếu A thì B (If A then B)
6	Nếu A hay B (If A or B)
7	A và B nhưng C (If A and B, C)
8	A và B nhưng C (If A and B but C)
9	A và B và C (A and B and C)
10	A hoặc B và C (A or B and C)

4.4 Linguistic Analysis Based on Grammar

By linguistic analysing, we can be deep understanding in text, we implemented in following steps. Step 1: Indentify sentence structure by Vietnamese grammar via rules and Step 2: identify clauses in sentence, where is the main clause, sub clause and which clause is showed the idea or meaning from user. We implemented 50 rules in step 1 (Table 11).

4.5 Extracting Features

A sentiment token is a word or a phrase that conveys sentiment. Given those sentiment words, a word token consists of a positive (negative) word and its part-of-speech tag. Sentiment tokens and sentiment scores are information extracted from the original dataset. They are also known as features, which will be used for sentiment categorization. In order to train the classifiers, each entry of training data needs to be transformed to a vector that contains those features, namely a feature vector. For the sentence-level categorization, a feature vector is formed based on a sentence (Fig. 2).

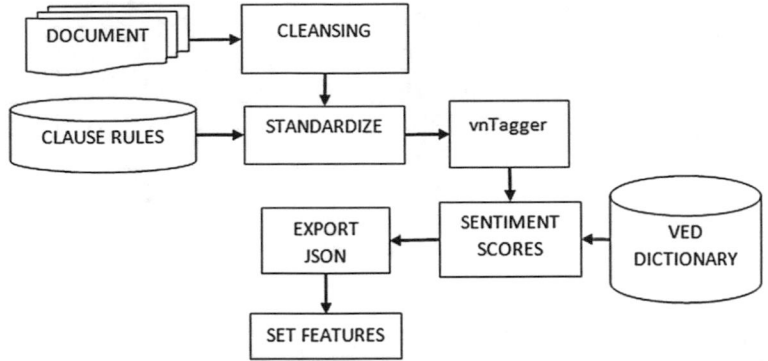

Fig. 2 Extracting features model

4.6 Sentiment Analysis

Evaluating the sentence has emotion or non-emotion based on features which were selected adaptation in Vietnamese language and Vietnamese emotional dictionary, and then sentiment classification algorithm has been processed to evaluate the emotion for sentence is in positive, negative or normal status based on support vector machine (SVM) classification method.

5 Experimental Models and Results

5.1 Training

As we know, emotional is extremely complicated. Hence to build a manageable data, we conducted collecting comments of user and labeled those each sentence in comment to analyze them. Data used in this paper is a set of product reviews collected from a famous technology site—tinhte.vn. We collected, in total, over 115.000 product reviews in which the products belong to 3 major categories in 400 topics: smart phone, watch, electronic device. Each sentence has subjective nature of every person. The first task is to classify which comment is emotional or non-emotional (also known as subjectivity classification) and the second task is to classify which comment is negative or positive (also known as sentiment classification).

In the next step, we divide manually the synthetic database into 2 parts: subjective and objective sentences. After that, the subjective sentences were classified manually into 2 parts: negative and positive sentence (Tables 12 and 13).

Table 12 Result of subjective manually classification

Number	Topic	Training data	
		Subjective sentences	Objective sentences
1	Smart phone	22,640	12,123
2	Watch	34,150	15,417
3	Electronic device	23,200	10,901
4	All	79,990	38,441

Table 13 Result of sentiment manually classification

Number	Topic	Training data	
		Positive sentences	Negative sentences
1	Smart phone	11,015	11,625
2	Watch	20,187	13,963
3	Electronic device	13,400	9800
4	All	44,602	35,388

5.2 Subjective Classification

This method uses 6 features to classify which sentence is emotional or non-emotional:

- Feature 1st: The amount of word in the sentence. It partly displays what the users want to express through the comments. If a lots of number of words are appeared, the user is really interested in this topic.
- Features 2nd, 3rd, 4th and 5th: The total of emotional value of noun, adjective, verb, adverb in the comments. The emotional value of a sentence depends on the type of word which was compared with the VED dictionary.
- Feature 6th: The total of emotional value of a sentence is basically total of 4 attributes that is 2nd, 3rd, 4th, 5th.
- Moreover, the emotional value of a sentence also depend on the type of the sentence. The emotional value of a sentence will be 0 point if this sentence is a condition or a question sentence.

Algorithm 1:

- Input: Sentence has been preprocessing, VED emotional dictionary.
- Output: Feature vectors
- Steps:
  - Count number of words in sentence
  - Find and calculate sum of emotional value of adjective in sentence. (2)
  - Find and calculate sum of emotional value of adverb in sentence. (3)
  - Find and calculate sum of emotional value of noun in sentence. (4)
  - Find and calculate sum of emotional value of verb in sentence. (5)

- Sum of emotional value of sentence = sum of all values in 2, 3, 4, 5
- If sentence is question or conditional sentence return 0
  Otherwise, return sum of emotional value of sentence.
- Return feature vector.

From the feature vector, we use SVM method to classify sentence into subjective (emotional) or objective class (non-emotional).

5.3   Sentiment Classification

After the subjective classification has been processed, we continued to apply the sentiment classification on these sentences. We proposed features consistent with Vietnamese language.

- Firstly, emotional value of a sentence depends on the emotional value of each emotional word or phrase. The most basic attributes inherited from subjective analysis. The summary of emotional value of a sentence is total in value of all features above.
- Secondly, emotional value of a sentence which depends on the emotional value of the intensification will be calculated by: Emotional value = value of intensification * value of emotional word. The total of these values will be the new value of the emotion after review intensification. In the absence of intensification in sentence, this value is the total value of all kinds of emotional words in a sentence.
- Thirdly, emotional value of a sentence also depends on the negative words in the sentence: "không" (no), "không có" (without), ... will be calculated by: Emotional value = (-1) * value of emotional word
- Fourthly, emotional value of a sentence which depends on the imperfect words: "nên" (should), "phải" (must have), "có thể" (maybe), ... will be calculated by: Emotional value = (0.5) * total value of all imperfect words in a sentence
- Fifthly, emotional value of a positive sentence: In fact, traditional Vietnamese culture, people avoid using negative words to express their opinions so that the positive words are commonly used. Hence, the emotional value of a positive word will be calculated by: Emotional value = $(1 + 0.5)$ * value of positive word
- Lastly, emotional value of a sentence which has a contrasting-linked word likes: "nhưng"(but),"tuy nhiên" (however), ... will be calculated by total of the emotional value of words that subtract the emotional value of the words before the contrasting − linked word by: Emotional value = Emotional value − total of emotional value of the words before the contrasting − linked word

Algorithm 2:

- Input: Sentence has been preprocessing and VED emotional dictionary.

- Output: Feature vectors
- Steps:

For each sentence from the data do:

- Find and calculate sum of emotional value of adjective in sentence. (2)
- Find and calculate sum of emotional value of adverb in sentence. (3)
- Find and calculate sum of emotional value of noun in sentence. (4)
- Find and calculate sum of emotional value of verb in sentence. (5)
- Sum of emotional value of sentence = sum of all values in 2, 3, 4, 5
- Find intensification words in the sentence and update the value of emotional:
  *Emotional value = value of intensification * value of emotional word*
- Find negative words in the sentence and update the value of emotional:
  *Emotional value = (−1) * value of negative word*
- Find imperfect words in the sentence and update the value of emotional:
  *Emotional value = (0.5) * total value of all imperfect words in a sentence*
- Find positive words in the sentence and update the value of emotional:
  *Emotional value = (1 + 0.5) * value of positive word*
- Find linked word in the sentence and update the value of emotional:
  Emotional value = Emotional value − total of emotional value of the words before the contrasting − linked word
- Return feature vector.

5.4 Result and Discussion

5.4.1 Randomize Test

We used 3 data sets from 3 categories smart phone, watch and electronic device for evaluation. Each data sets has 250 sentences which selected in random. By using our features which had consistent with Vietnamese language and classify based on SVM classification method and then we calculated the precision measure of subjective classification according to the algorithm in previous step. Accuracy measure is calculated as the proportion of the true sentence classification against all the sentences (Tables 14 and 15).

Table 14 Results of subjective classification

ID	Data set	Number sentences	Sentiment evaluation accuracy (%)
1	Smart phone	250	94.30
2	Watch	250	91.70
3	Electronic device	250	90.50
4	ALL	750	92.16

Table 15 Results of sentiment classification

ID	Data set	Number sentences	Sentiment evaluation accuracy (%)
1	Smart phone	250	83.50
2	Watch	250	82.70
3	Electronic device	250	80.80
4	ALL	750	82.33

Table 16 Results of subjective classification

ID	Smart phone	Watch	Electronic device
	Sentiment evaluation accuracy (%)	Sentiment evaluation accuracy (%)	Sentiment evaluation accuracy (%)
1	92.10	92.00	91.21
2	91.04	92.25	90.05
3	93.01	92.73	89.68
4	90.08	92.22	90.00
5	89.29	90.61	88.71
6	91.78	91.80	89.00
7	93.05	87.69	85.21
8	92.17	89.25	84.62
9	91.56	92.34	86.89
10	90.89	92.19	89.10
AVG	91.49	91.30	88.44

5.4.2   Cross Validation Test

We evaluated accuracy in average using tenfold cross validation. A tenfold cross validation is applied as follows: a data set is partitioned into 10 equal size subsets, each of which consists about 7000 sentences belong to positive class and negative class. Of the 10 subsets, a single subset is retained as the validation data for testing the classification model, and the remaining 9 subsets are used as training data. The cross-validation process is then repeated 10 times, with each of the 10 subsets used exactly once as the validation data. The 10 results from the folds are then averaged to produce a single estimation (Tables 16 and 17).

5.4.3   Component Test

In previous section, we mentioned that two important components affect to result in sentiment analysis including: Emotional dictionary, linguistic analysis based on grammar. We want to evaluate how to each component affect and role each component in the system, so we tested system with features when combine each component (Tables 18 and 19).

Table 17 Results of sentiment classification

ID	Smart phone	Watch	Electronic device
	Sentiment evaluation accuracy (%)	Sentiment evaluation accuracy (%)	Sentiment evaluation accuracy (%)
1	82.15	82.20	81.12
2	81.67	82.25	80.56
3	83.18	81.34	79.82
4	79.43	82.22	78.78
5	79.29	80.61	78.97
6	84.00	81.80	82.00
7	80.05	83.01	81.21
8	82.46	79.95	83.00
9	81.96	82.22	80.89
10	85.09	83.01	82.08
AVG	81.93	81.86	80.84

Table 18 Results of subjective classification

ID	Features	Accuracy (%)
1	VED Emotional dictionary	67.70
2	Linguistic analysis	80.40
3	VED Emotional dictionary + Linguistic analysis	92.68

Table 19 Results of sentiment classification

ID	Features	Accuracy (%)
1	VED Emotional dictionary	65.20
2	Linguistic analysis	71.20
3	VED Emotional dictionary + Linguistic analysis	83.50

In the results, we saw that the system improved accuracy than other system in Vietnamese for the same domain. In particular, with the randomize test, results of subjective classification is 94.30% accuracy and 83.50% accuracy for sentiment classification.

In the cross validation test, results of subjective classification is 91.49% accuracy in average and 81.93% accuracy in average for sentiment classification. From the results, we can confirmed that the system has very good performance and can be applied in online systems. Especially, the result in experiments is very stability because of features which were selected adaptation in Vietnamese language and Vietnamese emotional language has been built with more words consistent with the Vietnamese grammar based on spelling that people are using on social network.

When we saw the result from component test, we also confirmed that in Vietnamese language, linguistic analysis is very important for sentiment analysis, because weight in affection to result much more than emotional dictionary, this mean that when we analytic sentence in language, struct language and syntactic is the key to more understanding meaning in sentence. This point is going to deep consider when we try to improve the system can evaluate domain dependent implicit sentiment as it does not train on any domain specific data.

6 Conclusion

Sentiment detection has a wide variety of applications in information systems, including classifying reviews, summarizing review and other real time applications. In this paper, we proposed a framework for sentiment analysis based on combining lexicon-based and learning-based methods for sentiment analysis for product review and in Vietnamese language. We built Vietnamese emotional dictionary (VED) which contains 5 sub-dictionaries in manually, including noun, verb, adjective, and adverb dictionary and a special part emotional words which appear in Vietnamese in everyday life. In addition, we added some words to our dictionary to make consistent with the Vietnamese grammar and concise spelling that people are using on network. Combining lexicon-based and learning-based methods to be used in online product review. The experimental show that our system has very well performance and can be applied in online systems. In the future, we try to improve the system can evaluate domain dependent implicit sentiment as it does not train on any domain specific data.

References

1. Trinh, S., Nguyen, L., Vo, M., Do, P.: Lexicon-based sentiment analysis of Facebook comments in Vietnamese language. In: Studies in Computational Intelligence, pp. 263–276. Springer (2016)
2. Liu, B.: Sentiment analysis and opinion mining. In: Synthesis Lectures on Human Language Technologies, pp. 1–167. Morgan & Claypool Publishers (2008)
3. Aue, A., Gamon, M.: Customizing sentiment classifiers to new domains: a case study. Presented at the Recent Advances in Natural Language Processing (RANLP), Borovets, Bulgaria (2005)
4. Andreevskaia, A., Bergler, S.: ClaC CLaC-NB: Knowledge-based and corpus-based approaches to sentiment tagging. In: Proceedings of the 4th International Workshop on Semantic Evaluations. Association for Computational Linguistics, Prague, Czech Republic (2007)
5. Vinodhini, G., Chandrasekaran, R.M.: Sentiment analysis and opinion mining: a survey. Int. J. Adv. Res. Comput. Sci. Softw. Eng. 2(6) (2012)
6. Turney, P.D.: Thumbs up or thumbs down? Semantic orientation applied to unsupervised classification of reviews. In: Proceedings of the 40th Annual Meeting of the Association for Computational Linguistics (ACL), pp. 417–424 (2002)
7. Taboada, M., Brooks, J., Tofiloski, M., Voll, K., Stede, M.: Lexicon-based methods for sentiment analysis. Comput. Linguist. 37(2), 267–307 (2011)

8. Melville, P., Gryc, W., Lawrence, R.D.: Sentiment analysis of blogs by combining lexical knowledge with text classification. In: Proceedings of 15th ACM SIGKDD International Conference on Knowledge Discovery and Data Mining (2011)
9. Ding, X., Liu, B., Yu, P.S.: A holistic Lexicon-based approach to opinion mining. In: Proceedings of the International Conference on Web Search and Web Data Mining, pp. 231–240. ACM (2008)
10. Hatzivassiloglou, V., McKeown, K.R.: Predicting the semantic orientation of adjectives. In: Proceedings of the eighth conference on European chapter of the Association for Computational Linguistics, pp. 174–181. Association for Computational Linguistics, Madrid, Spain (1997)
11. Polanyi, L., Zaenen, A.: Contextual valence shifters. In: Computing Attitude and Affect in Text: Theory and Applications, pp. 1–10. Springer, Dordrecht (2006)
12. Kennedy, A., Inkpen, D.: Sentiment classification of movie reviews using contextual valence shifters. Comput. Intell. **22**(2), 110–125 (2006)
13. Nguyen, H.N., Van Le, T., Le, H.S., Pham, T.V.: Domain specific sentiment dictionary for opinion mining of Vietnamese text. In: The 8th International Workshop, MIWAI 2014, Bangalore, India (2014)
14. Duyen, N.T., Bach, N.X., Phuong, T.M.: An empirical study on sentiment analysis for Vietnamese. In: International Conference on Advanced Technologies for Communications (2014)
15. Duy, N.N.: Document summarization based on sentiment classification. Master thesis in computer science (Vietnamese), University of Technology Hochiminh City (2014)
16. Phu, V.N., Tuoi, P.T.: Sentiment classification using enhanced contextual valence shifters. In: Proceedings of International Conference on Asian Language Processing, Malaysia (2014)
17. Pang, B., Lee, L.: A sentimental education: sentiment analysis using subjectivity summarization based on minimum cuts In: Proceedings of the 42nd Annual Meeting on Association for Computational Linguistics, ACL '04. Association for Computational Linguistics, USA (2004)
18. Characteristics of Vietnamese language. http://en.wikipedia.org/wiki/Vietnamese_alphabet

Reducing Misclassification of True Defects in Defect Classification of Electronic Board

Tokiko Shiina, Yuji Iwahori, Yohei Takada, Boonserm Kijsirikul and M.K. Bhuyan

Abstract This paper proposes a method to discriminate the defect on the electronic board and the foreign matter attached to the circuit. The purpose of this paper is to reduce the misclassification of the true defect to the pseudo defect in the automatic classification approach. Proposed method improves the classification accuracy under the multiple illuminating conditions. Each result of multiple classifiers is used for the voting process and incorrect classification is reduced by introducing the classification into three classes of true defect, pseudo detect, and difficult defect. The approach evaluates the correct ratio for the circuit board image based on actual defects are included and the effectiveness of the proposed approach is confirmed via the experiment.

Keywords SVM · Multiple classes · Defect candidate region · Defect classification

T. Shiina (✉) · Y. Iwahori · Y. Takada
Graduate School of Engineering, Chubu University, 1200, Matsumoto-cho,
Kasugai 487-8501, Japan
e-mail: shiina@cvl.cs.chubu.ac.jp

Y. Iwahori
e-mail: iwahori@cs.chubu.ac.jp

Y. Takada
e-mail: ytakada@cvl.cs.chubu.ac.jp

B. Kijsirikul
Department of Electronics, Chulalongkorn University,
Bangkok 20330, Thailand
e-mail: Boonserm.K@chula.ac.th

M.K. Bhuyan
Department of Electronics & Electrical Engineering, IIT Guwahati,
Guwahati 781039, India
e-mail: mkb@iitg.ernet.in

© Springer International Publishing AG 2018
R. Lee (ed.), *Computer and Information Science*, Studies in Computational
Intelligence 719, DOI 10.1007/978-3-319-60170-0_6

1 Introduction

Printed circuit board (PCB) is crucial part of electronic device; it needs to be properly investigated before get launched. Automatic inspection systems are used for this purpose, but due to more complexity in circuits, PCB inspections are now more problematic. This problem leads to new challenges in developing advanced automatic visual inspection systems for PCB.

Automatic optical inspection (AOI) has been commonly used to inspect defects in printed circuit board during the manufacturing process. An AOI system generally uses methods which detects the defects by scanning the PCB board and analyzing it. AOI uses methods such as local feature matching, image skeletonization, and morphological image comparison to detect defects and has been very successful in detecting defects in most of the cases, but production problems such as oxidation, dust, contamination, and poor reflecting materials leads to most inevitable false alarms. Reducing the false alarms is the concern of this paper.

Two kinds of defect exist in the electronic circuit board, one of which is the true defect and the other of which is the pseudo defect. True defects are classified into disconnection, connection, projection, and crack. These defects have problem to pass them to the market. Pseudo defects are classified into oxidation and dust, and pseudo defects are still safe to the market. Misclassification of pseudo defect to true defect means discarding the normal electronic board, which wastes the production cost. Inversely, misclassification of true defect to pseudo defect means manufacturing problem to the market.

Both cases are problems, but the required condition is to reduce the misclassification of true defect to pseudo defect as much as possible, and its ideal condition should have 0 misclassification of true defect to pseudo defect. Required condition is misclassification of pseudo defect to true defect should be less than or equal to 30 % under the above condition.

There are previous approaches for detecting defects of electronic circuit board such as [1, 2]. Maeda et al. [1] uses infrared light image to detect the defect by testing electrically, while Numada et al. [2] extracts the global features using the interest point and its surrounding points and detects the defect using Mahalanobis distance.

Papers [3–7] have also been proposed. Rau et al. [3] detects the defect using subtraction of the original CAD data and test image and classifies the defect based on the detected shape information. Roh et al. [4] detects the defect using the logical operation of binary images of test image and reference image and classifies the defect using neural network. Features to classify the defects in these papers are obtained from binary image without any color information and classifies kinds of true defects into lack or projection, while pseudo defect is not treated in these papers.

Futamura et al. [5] performs two classes classification for true or false defects. One SVM [8] is used, and there is some possibility that the identification boundary is affected when noise is included in the learning data. Hagi et al. [6] treats the electronic board image with 918 × 684 pixels which was taken under a single light.

Random sampling is used to construct the subset, and SVM is learned for each subset. Weighted majority voting is applied from the output of each SVM to perform two classes classification. The image size used is small with 256×256 and the defect candidate region becomes very small. This causes problem to reduce the accuracy for classification since the observation gives very small difference for features under a single lighting condition.

Takada et al. [7] classifies test samples using SVM under the condition that two images are taken under the different lighting conditions without using reference image. However, this approach obtains the defect candidate region with the region segmentation for a test image, and there is a problem that it is still difficult to correspond to the defect with the region consisting of unclear segmentation. These approaches aim to increase the total accuracy but do not challenge to decrease the number of misclassification for the true defects.

This paper proposes a new approach for the defect classification of electronic board with considering the accuracy for the true defects using two test images taken under the different lighting conditions. Hagi et al. [6] classifies the defects under a single illuminating condition, and this paper improves the classification accuracy under multiple illuminating conditions. While Hagi et al. [6] uses the voting process from the output of multiple classifiers and performs two classes classification between true defect and foreign matter, the proposed method further reduces the misclassification by introducing three classes classification including difficult class. The effectiveness of the proposed approach was evaluated for the probability of true positive and true negative for the defect classification.

2 Lighting Condition and Kinds of Defect

Test image used for this research is the detected defect by AOI, which was taken again by the device with the human eye check called "verification device." Verification device takes two kinds of images under two different lighting conditions. Image taken with large angle lighting is named coaxial main, and image taken with both side lighting and large angle lighting is named side main in this paper. Coaxial main illuminates the board from the vertical direction as shown in Fig. 1a while side main illuminates board from side directions in addition to the vertical direction as shown in Fig. 1b.

Coaxial main is effective to sticking foreign matter, discoloration, or lack. Side main is effective to the projection and appearance of foreign matter. Examples taken under each lighting condition are shown in Fig. 2.

Lack shown in Fig. 2a and projection shown in Fig. 2b are the true defects. Color of the region of short and projection has the same as that of lead line part in both of coaxial main and side main. Color of the region of disconnection and lack has the same as that of base part in both of coaxial main and side main. Image taken from side main tends to shine in the part of edge of defect region.

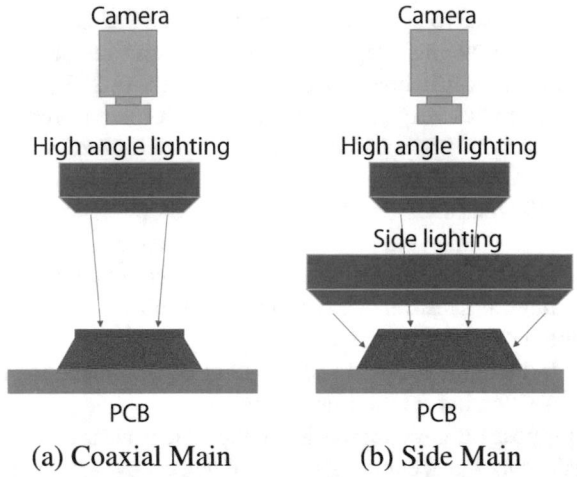

Fig. 1 Lighting condition

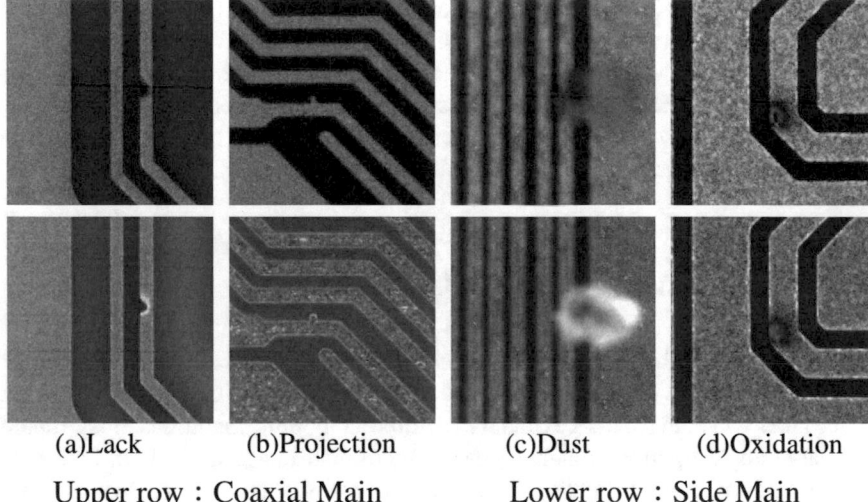

| (a)Lack   (b)Projection | (c)Dust   (d)Oxidation |
| Upper row : Coaxial Main | Lower row : Side Main |

Fig. 2 Kind of defect

Dust shown in Fig. 2c and oxidation shown in Fig. 2d represent pseudo defects. Adhesion of dust etc. gives the same color as that of base part in the image of coaxial main but shines in the image of side main. Discoloration changes color in the defect part in both images of coaxial main and side main, while it shows the different color in lead line part and base part.

3　Methodology

Defect candidate region is detected from a test image and a reference image. Features are extracted from the defect candidate region, and combination of effective features is determined by the forward sequential selection. This processing is done for each of classifier (SVM) constructed using random sampling with different combination of features. Selected features are learned by each SVM, and output obtained from each classifier is used with majority voting to identify true defect or pseudo defect.

3.1　Detection of Defect Candidate Region

Reference image is prepared for each test image, and defect candidate region is detected to extract features. Test image of coaxial main sometimes gives no difference of color between defect candidate region and base part when color of adhesion is close to that of base part. So, the detection is performed to the test image of side main. Features are extracted from the detected defect candidate region of both images of coaxial main and side main (Figs. 3 and 4).

　　Difference image is made from test image of side main and reference image, and its difference image is binarized with the threshold for each component of RGB. Here, threshold is determined from the discriminant analysis method, and binarized image of each component is generated. Taking logical OR of binary images of each component then closing and opening processing are applied to remove the noise and mask image is generated (Figs. 5 and 6).

Fig. 3　Test image

Fig. 4　Reference image

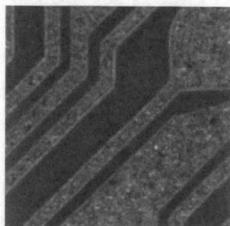

Fig. 5   Difference image

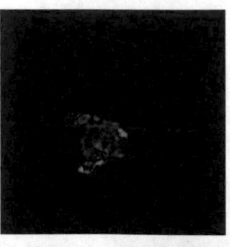

Fig. 6   Mask image

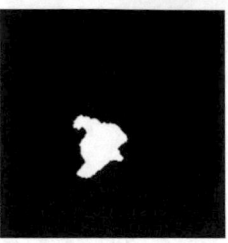

3.2   Feature Extraction

Features are extracted from the defect candidate region. Color information and shape information are extracted as features and extracted features are used for learning and classification by SVM.

3.2.1   Color Information

Color information is extracted from each of coaxial main image and side main image. Color information consists of a total of 192 kinds which are (1) Maximum, (2) Minimum, (3) Mean, (4) Proportion of High Value, (5) Ratio of lead line region and candidate region and lead line, (6) Ratio of base part and candidate region, (7) Position difference between center of gravity of value and the maximum value, (8) Variance, (9) Standard Deviation, (10) Kurtosis, (11) Skewness, (12) Entropy, (13) Difference between maximum and minimum, (14) Mode, (15) Median, and (16) Correlation between test image and reference image. These features are obtained from a total of 192 kinds for each of RGB and HSV.

3.2.2   Shape Information

Shape information is obtained from the defect candidate region of mask image. Shape information consists of a total of 8 kinds including (1) Area, (2) Perimeter, (3) Size of x-direction, (4) Size of y-direction, (5) Aspect Ratio, (6) Diagonal Length, (7) Complexity, and (8) Circular Degree.

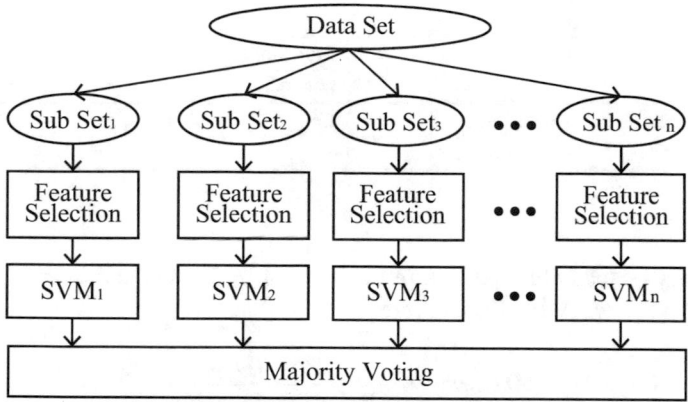

Fig. 7 Constructed classifiers

3.3 Normalization of Features

See Hagi et al. [6] for normalization of features.

3.4 Classifier

Subset is made using random sampling from the learning data set D and n subsets $D_1, \ldots, D_n$ is constructed. Each classifier is constructed for each subset. Introducing random sampling-based learning aims robust learning to the noise. Features with any number of dimension are randomly selected to be used for classifier. Feed forward selection is applied to the selected features and SVM learning is done using the effective features selected for the discrimination. Test data are input to the classifier which has been constructed for each subset. The corresponding output from each classifier is voted, and the final classification is performed with majority voting processing. Constructed classifiers are shown in Fig. 7. See Hagi et al. [6] for making subsets.

3.5 Evaluation for Feature Selection

Feature selection is applied using forward sequential selection from randomly selected features to reduce the misclassification of true defect. Let true defect be labeled as Positive and let pseudo defect be labeled as Negative. Confusion matrix is shown in Table 1.

Table 1 Confusion matrix

		True value	
		True defect	Pseudo defect
Estimation result	True defect	TP	FP
	Pseudo defect	FN	TN

Accuracy of total classification result is given by Eq. (1) and Correct Ratio for true defect is given by Eq. (2) in Table 1.

$$Accuracy = \frac{TP + TN}{TP + FP + TN + FN} \tag{1}$$

$$Recall = \frac{TP}{TP + FN} \tag{2}$$

Recall is most important index in defect classification. Evaluation with high correct ratio for true defect and with high accuracy for whole classification under this condition is given by Eq. (3).

$$Evaluation\ Value = Accuracy \cdot Recall \tag{3}$$

3.6 Majority Voting Processing

Majority voting processing is applied from results of each SVM constructed for each subset. The judgement result is classified into true defect, pseudo defect, or difficult classification. When the result of each SVM is judged as true defect, point +1 is added, while when it is judged as pseudo defect, point −1 is added and total point calculated with this voting processing. When total point is positive, judgement is taken as true defect and when total point is negative, judgement is taken as pseudo defect. Here, absolute value of this total score takes less than some threshold, judgement is taken as defect of difficult classification. Let this threshold be th, then judgement is done as

$$Result = \begin{cases} True\ Defect & (Score > th) \\ Pseudo\ Defect & (Score < -th) \\ Difficult & (otherwise) \end{cases} \tag{4}$$

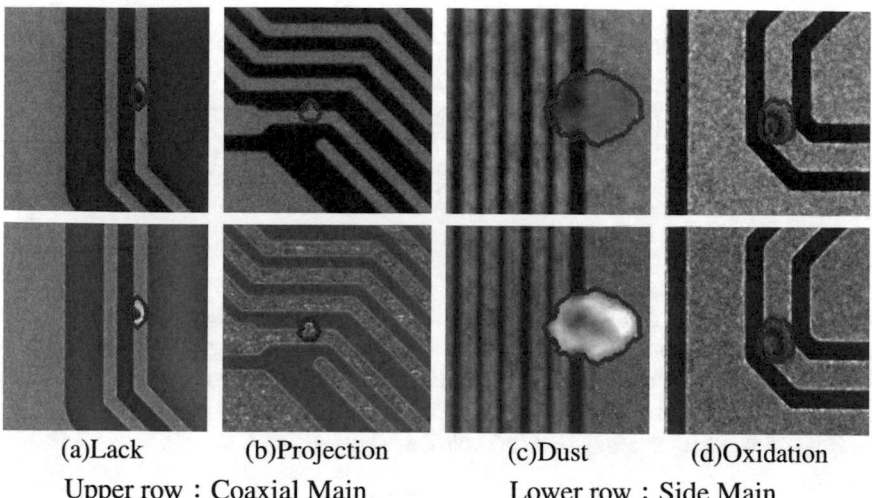

| (a)Lack | (b)Projection | (c)Dust | (d)Oxidation |

Upper row : Coaxial Main   Lower row : Side Main

Fig. 8 Defect detection

4 Experiments

Actual electronic board images are used for defect classification. Features are extracted from defect candidate region. Images are taken by the verification device for the defect detected by AOI. Image size is 256×256, and two images are used under two lighting conditions.

4.1 Defect Detection

Defect candidate region was detected using test image and reference image. Figure 8a shows the detected result of lack, Fig. 8b shows that of projection, Fig. 8c shows that of dust, and Fig. 8d shows that of oxidation. It is shown that defect candidate region is detected with high accuracy for each defect from Fig. 8.

4.2 Defect Classification

Method [6] and proposed approach are performed for comparison. Dataset consists of 85 true defect images and 72 pseudo defect images, i.e., a total of 157 images. Evaluation was done by tenfold cross validation [9]. RBF kernel was used for SVM. Number of subset was 30 and number of learning data of SVM for each subset was 50. Parameter C of SVM is taken from 1 to 100 (with step 1) while parameter σ

Table 2 Confusion matrix

		True value	
		True defect	Pseudo defect
Estimation result	True defect	TP	FP
	Pseudo defect	FN	TN
	Difficult	DP	DN

was taken for $0 < \sigma \leq 1.0$ (with step 0.0001), and these parameters were randomly obtained and used for each SVM. Difference of point to be judged as difficult classification class was set to be less than or equal to 15% of number of subsets. Confusion matrix of classification result is shown in Table 2.

Equations used for evaluation in Table 2 are below.

$$Accuracy = \frac{TP + TN}{TP + FP + TN + FN} \tag{5}$$

$$True\ Positive\ Rate = \frac{TP}{TP + FN} \tag{6}$$

$$True\ Negative\ Rate = \frac{TN}{FP + TN} \tag{7}$$

4.3 Lighting Condition

Classification performance was evaluated under only coaxial main (Hagi et al. [6]), only side main, and both of them using two images. Classification result is shown in Table 3.

Case of Classification under only coaxial main gives highest correct ratio for true defect, but this does not satisfy the classification accuracy with more than 70% for pseudo defect and has many images for difficult classification. While case of two images under coaxial main and side main satisfies the classification accuracy with more than 70% for pseudo defect, this case gives higher correct ratio for true defect and has lower number of images for difficult classification. Case with two images are used gives effective classification.

4.4 Evaluation of Feature Selection

Evaluation with forward sequential selection method was compared with the proposed approach for total accuracy (Hagi et al. [6]) and correct ratio for true defect.

Table 3 Classification result for each lighting condition

(a) Classification result

	True defect			Pseudo defect		
	Correct	Incorrect	Difficult	Correct	Incorrect	Difficult
Coaxial main	75	0	10	19	15	38
Side main	65	4	16	40	15	17
Proposed	74	1	10	41	11	20

(b) Accuracy (%)

	True	Pseudo	Total
Coaxial main	100.00	55.88	86.24
Side main	94.20	72.73	84.68
Proposed	98.67	78.85	90.55

Table 4 Classification result for each evaluation value

(a) Classification result

Evaluation value	True defect			Pseudo defect		
	Correct	Incorrect	Difficult	Correct	Incorrect	Difficult
Eq. (1)	64	11	10	58	5	9
Eq. (2)	81	0	4	1	44	27
Proposed	74	1	10	41	11	20

(b) Accuracy (%)

Evaluation value	True	Pseudo	Total
Eq. (1)	85.33	92.06	88.41
Eq. (2)	100.00	2.22	65.08
Proposed	98.67	78.85	90.55

Table 4 shows that the correct ratio of pseudo defect gives the highest value and that of the true defect gives the lowest value when total accuracy is evaluated. While, the number of misclassification of true defect is 0, instead most pseudo defects are judged as true defects. While proposed approach gives the highest correct ratio of true defect with 98.67% and that of pseudo defect gives satisfied level of request, it is confirmed that the proposed approach is useful.

4.5   Voting Processing

Case of two classes classification into true defect or pseudo defect with voting processing (Hagi et al. [6]) and case of three classes classification by adding difficult classes are compared here. Case without difficult class judges as

$$Result = \begin{cases} True\ Defect & (Score \geq 0) \\ Pseudo\ Defect & (Score < 0) \end{cases} \tag{8}$$

Table 5 shows the result of two classes classification without difficult class, while Table 6 shows that of three classes classification with difficult class. Here, accuracy of three classes classification is calculated by removing the images judged into a difficult class.

It is confirmed that Tables 5 and 6 shows that case with adding difficult class gives higher accuracy and proposed approach is useful.

Table 5 Without difficult class

(a) Classification result

True defect		Pseudo defect	
Correct	Incorrect	Correct	Incorrect
81	4	49	23

(b) Accuracy (%)

True	Pseudo	Total
95.29	68.06	82.80

Table 6 With difficult class

(a) Classification result

True defect			Pseudo defect		
Correct	Incorrect	Difficult	Correct	Incorrect	Difficult
74	1	10	41	11	20

(b) Accuracy (%)

True	Pseudo	Total
98.67	78.85	90.55

4.6 Examples of Misclassification and Difficult Classification

Images of misclassification and difficult classification by the proposed approach are shown here. Figures 9 and 10 show true defects, while Figs. 11 and 12 show pseudo defects.

Examples of true defects in misclassification or difficult classification are shown in Fig. 9 and Fig. 10, respectively. These are because color of adhesion was closed to that around lack part or adhesion was attached at bottom of the short part. Examples of pseudo defects in misclassification or difficult classification are shown in Fig. 11 and Fig. 12, respectively. There are because the defect candidate region is small or color of adhesion was closed to that of base part and this was judged as short.

4.7 Effects of Threshold to Classification Result

Rate of misclassification and that of difficult class was calculated by changing the threshold value. In Table 2, Eq. (9) represents error rate of total misclassification, Eq. (10) represents that of true defect, Eq. (11) represents that of pseudo defect, Eq. (12) represents difficult rate of total classification, Eq. (13) represents that of true defect, Eq. (14) represents that of pseudo defect.

Fig. 9 Misclassification of true defect (Lack)

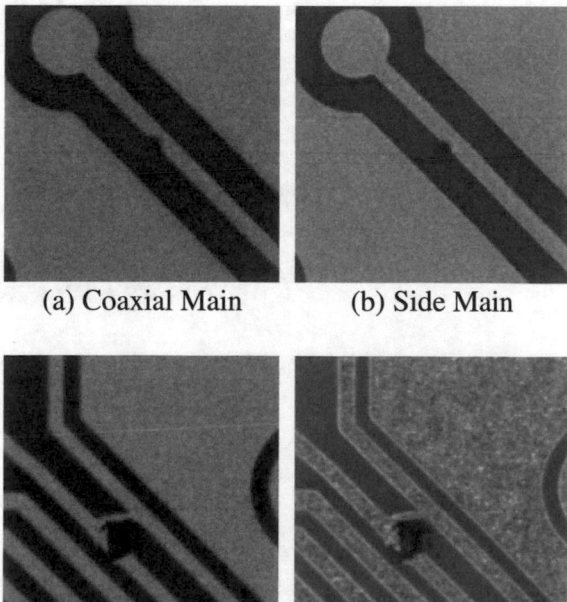

(a) Coaxial Main   (b) Side Main

Fig. 10 True defect in difficult classification (Short)

(a) Coaxial Main   (b) Side Main

Fig. 11 Misclassification of pseudo defect (Adhesion)

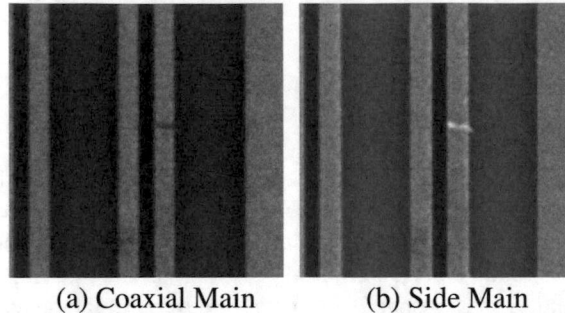

(a) Coaxial Main   (b) Side Main

Fig. 12 Pseudo defect in difficult classification (Adhesion)

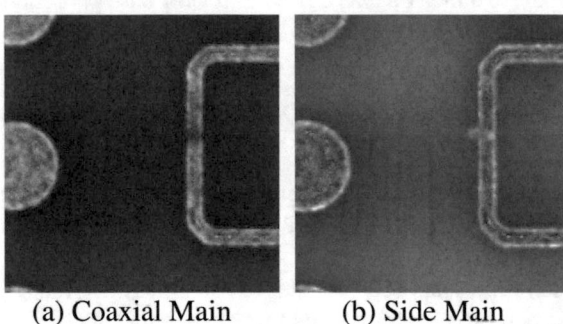

(a) Coaxial Main   (b) Side Main

$$Error\,Rate = \frac{FP + FN}{TP + FP + TN + FN} \tag{9}$$

$$True\,Defect\,Error\,Rate = \frac{FP}{TP + FP} \tag{10}$$

$$Pseudo\,Defect\,Error\,Rate = \frac{FN}{TN + FN} \tag{11}$$

$$Difficulty\,Rate = \frac{DP + DN}{TP + FP + DP + TN + FN + DN} \tag{12}$$

$$True\,Defect\,Difficulty\,Rate = \frac{DP}{TP + FN + DP} \tag{13}$$

$$Pseudo\,Defect\,Difficulty\,Rate = \frac{DN}{TN + FP + DN} \tag{14}$$

Figure 13 and Fig. 14 show the rate of misclassification and that of difficult classification, respectively.

Figure 13 shows that the number of misclassification decreases as the threshold value increases, and that error rate of true defect becomes 0% when threshold value becomes 10. Error rate of pseudo defect is 11.11%, and this satisfies the

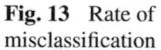

Fig. 13 Rate of misclassification

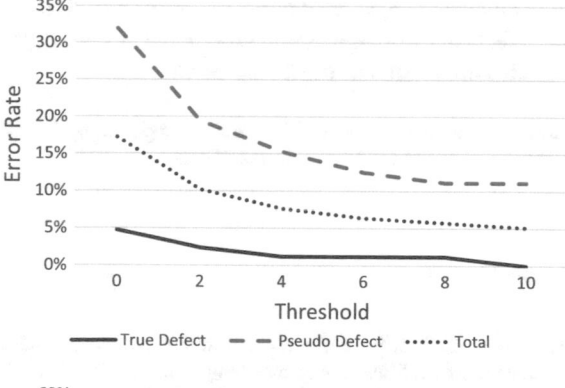

Fig. 14 Rate of difficult classification

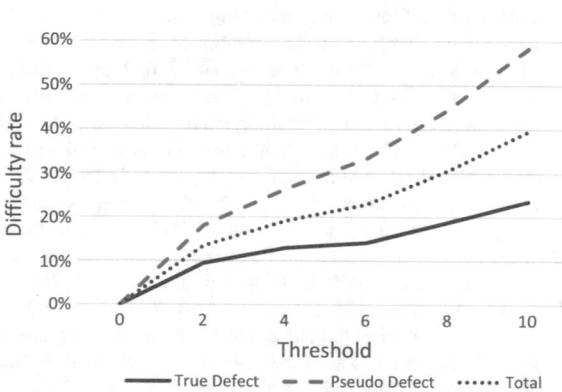

request under this condition. Figure 14 shows that the rate of difficult classification increases according to increasing threshold and error rate of total classification becomes 39.49% when error rate of true defect becomes 0% with threshold value 10. Figures 13 and 14 shows the relation between the rate of difficult classification, and its accuracy becomes trade-off each other. It is necessary to determine the threshold for the desired accuracy based on the product.

5 Conclusion

This paper proposed a new approach to classify the defect into three classes using two images under different lighting conditions.

Using two images with side main in addition to coaxial main could give the higher performance of classification than the single lighting condition. Weight is added to evaluate the accuracy of true defect and voting processing was applied to classify the defect into three classes. This could decrease the misclassification of true defect and satisfy the required condition for the correct rate of pseudo defect.

Remained subject is to perform 0 error rate of true defect and to decrease number of defects of difficult classification. Further classification is remained for the defects which were judged as difficult classes

Acknowledgements Iwahori's research is supported by JSPS Grant-in-Aid for Scientific Research (C) (26330210) and Chubu University Grant.

References

1. Maeda, S., Ono, M., Kubota, H., Nakatani, M.: Precise detection of short circuit defect on TFT substrate by infrared image matching. J. IEICE **J80-D-II CD-II**(9), 2333–2344 (1997)
2. Numada, M., Koshimizu, H.: A method for detecting globally distributed defects by using learning with Mahalanobis distance. In: ViEW2007, pp. 9–13 (2007)
3. Rau, H., Wu, C.-H.: Automatic optical inspection for detecting defects on printed circuit board inner layers. Int. J. Adv. Manuf. Technol. **25**(9–10), 940–946 (2005)
4. Roh, B., Yoon, C., Ryu, Y., Choonsuk, O.: A neural network approach to defect classification on printed circuit boards. JSPE **67**(10), 1621–1626 (2001)
5. Futamura, K., Iwahori, Y., Fukui, S., Kawanaka, H.: Defect classification of electronic board using SVM and table matching. In: MIRU2009, IS1-2, pp. 420–427 (2009)
6. Hagi, H., Iwahori, Y., Fukui, S., Adachi, Y., Bhuyan, M.K.: Defect classification of electronic circuit board using SVM based on random sampling. Proc. Comput. Sci. **35**, 1210–1218 (2014)
7. Takada, Y., Inoue, H., Shiina, T., Iwahori, Y.: Defect classification of electronic board using two images with changing lighting condition. In: C1-5, Shape/Object Recognition, Tokai Section Joint Conference on Electrical, Electronics, Information, and Related Engineering (2016)
8. Vapnik, V.N.: Statistical Learning Theory. Wiley (1998)
9. Efron, B.: Estimating the error rate of a prediction rule: improvement on cross-validation. J. Am. Stat. Assoc. **78**(382), 316–331 (1983)

Virtual Prototyping Platform for Multiprocessor System-on-Chip Hardware/Software Co-design and Co-verification

Arya Wicaksana and Tang Chong Ming

Abstract This paper describes the implementation of a virtual prototyping platform to address the ever-challenging multiprocessor system-on-chip (MPSoC) hardware/software co-design and co-verification requirements. The increasingly popular deployment of MPSoC brings complexity to system modeling, design, and verification. Fiercely competitive business environment makes it absolutely critical to rein in time-to-market and chip fabrication costs. The holy grail is to be able to verify the hardware design and synthesize to the gate level for physical layout, at the same time carry out software development for the hardware design using the same system models and verification platforms. One approach is to raise the abstraction level of system design and verification to ESL. In this paper, a virtual prototyping platform is built using SystemC with transaction-level modeling (TLM) and the open virtual platforms (OVP) processor model with instruction set simulator (ISS). As a demonstration of concept and feasibility, the virtual platform prototypes a 128-bit advanced encryption standard (AES) Cryptosystem MPSoC. The supporting subsystems and environment are also modeled, for example the system peripherals, the network-based interconnect scheme or Network-on-Chip (NoC), system firmware, the interrupt service handling, and driver. The virtual platform is scalable up to but not limited to twelve processing elements and configurable to the extent of the OVPs generic memory models (RAM and ROM) addresses and sizes, simulation parameters and debugging and tracing options.

Keywords Virtual prototyping platform · Transaction-level modeling · Multiprocessor system-on-chip · Hardware/Software co-design · Hardware/Software co-verification

A. Wicaksana (✉)
Department of Computer Science, Universitas Multimedia Nusantara,
Tangerang, Indonesia
e-mail: arya.wicaksana@umn.ac.id

T.C. Ming
Department of Electronic Engineering, Universiti Tunku Abdul Rahman,
Kampar, Malaysia
e-mail: tangcm@utar.edu.my

© Springer International Publishing AG 2018
R. Lee (ed.), *Computer and Information Science*, Studies in Computational
Intelligence 719, DOI 10.1007/978-3-319-60170-0_7

1 Introduction

The over increasing popular deployment of multiprocessor system-on-chip (MPSoC) today brings two major challenges: to build hardware that the software designers could use effortlessly and to develop system software which could completely utilizes the hardware potentials. The presence of many processors with their attending subsystems sets new challenges in hardware and software design. In today's extremely costly advanced technology nodes and competitive market environment, this ability to co-design and co-verify hardware and software concurrently can very easily be a matter of survival for the product design and development entity of an MPSoC. Shorter product life cycles, sky-rocketing cost, heightened time-to-market, explosive complexity, and greater competitive environment make it no longer acceptable for the software development to proceed after the hardware prototypes are made available as it used to be. They have to start concurrently. To ensure no inconsistency, the software will have to be developed using the same models and testbenches used in the synthesis and verification of the hardware design. The MPSoC designers and engineers inevitably have to face up to the new challenges that has turned into a common reality today as described in [14].

In MPSoC hardware design, new interprocessor communication scheme based on terrestrial telecommunication structure had become common place. This shift to network-on-chip (NoC) implementation was necessary as bus-based interconnect schemes were no longer able to manage multiple processors communication efficiently on the silicon level [4]. The main challenge in hardware design today is to be able to develop a hardware that is easy to use by software that meets consumer specification while satisfying market constraints e.g., reducing the cost, footprint, or power utilization. The exponential growth of the hardware and the semiconductor industries consequently brought new challenges to the software design league that never existed before. Software design complexity for multiple processors system is many times that for a single processor system. In the early days, system-on-chip (SoC) system software could only be tested after the hardware prototype was available. This was due to the limitation of design methodology and tools for both hardware and software. Thus, new methodologies and tools must be developed to enable the hardware and software designers to perform co-design and co-verification of the same design specification documents, models, and verification platforms.

The electronic system-level (ESL) design methodology as described in [3] proposed a virtual prototyping platform for solving MPSoC hardware/software co-design and co-verification problems. The virtual prototyping platform is categorized into several types based on the use cases. Each of the use cases targets specific purpose: early software development, architectural exploration, and verification. Based on the use cases, the abstraction level of the virtual prototyping platform is defined. There are three views: programmer's view, architectural's view, and verification view. Early software development will require faster simulation speed of the virtual prototyping platform to allow software engineers to boot firmware or even operating system (OS) within reasonable amount of time. This specific need obviously requires high-speed models which implies that the models do not carry much details implemented on them, as long as the models carry all of the functionalities that are required from the programmers view standpoint. On the other hand, for architecture exploration and verification purpose, high accuracy models are more favorable to provide estimates that can be used for exploring another architecture solutions and even more accuracy for doing verification in high-level.

In general, virtual platforms can be built using high-level languages e.g., SpeC, SystemC, and SystemVerilog along with the use of transaction-level modeling (TLM) for faster simulation speed. High-level reusable models of predefined standard components such as processors and memories are provided by open virtual platform (OVP) for ease of design. Other than that, high-level abstraction models that are suitable for fast high-level exploration at early design stages are developed using TLM with loosely timed (LT) coding style. However, the more accurate models which produce most precise results can be built using TLM with approximately timed (AT) coding style or without TLM in the cycle accurate (CA) abstraction level.

The objective of this paper is to discuss a general reference virtual prototyping platform for dealing with MPSoC hardware/software co-design and co-verification. We describe the specific characteristics of a virtual prototyping platform, and the features that may drive the adoption of virtual prototyping platform by both hardware/software co-design and hardware/software co-verification. Finally, we substantiate the discussion by reporting our experience in the AES-128 Cryptosystem MPSoC project, which is a proof-of-concept deployment of a virtual prototyping platform in dealing with MPSoC hardware/software co-design and co-verification. In this regard, we describe the technical solutions adopted for the realization of the virtual prototyping platform and report some of the measurements from it.

The rest of the paper is organized as follows. Section 2 overviews the technologies that are commonly associated with the hardware/software co-design and co-verification challenges of an MPSoC and that can be enabled by the deployment of a virtual prototyping platform. Section 3 provides a general overview of the system architecture for a virtual prototyping platform. More in detail, this section describes the TLM-LT approach for the realization of the virtual prototyping platform and also the hardware models in high-level of abstraction, with the related functionalities and communication protocols. Finally, Sect. 4 presents the AES-128 Cryptosystem MPSoC project, which exemplifies a possible implementation of a virtual prototyping platform and provides examples of the type of results that helps MPSoC engineers with the hardware/software co-design and co-verification.

2 Virtual Prototyping Platform

A virtual prototyping platform is basically a piece of software that mimics the true functionality of a complete system. In this case, it is an MPSoC. This way, the software engineers could perform early software development based on the virtual prototyping platform before the hardware is available. The usage is not limited only to

Table 1 Model abstraction levels

Level	Abstraction	Common names
Highest	Behavioral	Programmer's view (PV), untimed (UT)
Higher	Timed	PVT (Programmer's view + timing), loosely timed (LT), approximately timed (AT), cycle approximate
Lower	Cycle accurate	CA, clock accurate
Lowest	Implementation accurate	RTL, design simulation model (DSM)

that extent. Architectural exploration, functional verification, and estimated timing analysis could also be performed using the virtual prototyping platform. Each of the use cases mentioned in [5] is best achieved by building the virtual prototyping platform at specific abstraction level. The hardware engineers need implementation-accurate models to validate their designs. Meanwhile, the software engineers can get by with high-level behavioral models. The abstraction levels are described in Table 1.

The PV models [2, 10] provide virtual platform models with simulation speeds of ranges between 100 and 500 MIPS (million instructions per second). The characteristic of this platform is functionally accurate and executes really fast within the mentioned ranges. The reason why the simulation speed is the fastest amongst all is that there is no timing information at all (untimed) within the models. This allows the simulation speeds to be faster than more timing accurate models. Thus, these PV models are sufficient enough for software developer starting early software development [8]. More timing accurate models of these PV models could be built and generally called PV+T (programmers view + timing). Since time penalties increase with timing accuracy, designers usually start from a full LT model (untimed). As the design flow continues, modules with some time accuracy are added, [8] either via decomposition or through refinement. For the embedded software developer working with the processor architect, a modification requiring a change of the instruction set was almost immediate [6]: a new instruction set simulator (ISS) was generated and the embedded software could run very rapidly on the new ISS. The reason was that the processor was modeled in C as a functional model, and some wrapper code that represented the interface and communication to the processor peripherals.

The development of these abstract models is enabled by the design modeling language: SystemC [6]. SystemC, used as a vehicle to provide the TLM abstraction, has proven to be the key to the fairly fast deployment of this methodology [6]. There was no issue of proprietary language support by only one CAD (computer-aided design) vendor or university. There was also no issue of making a purchase decision by the design manager for yet another costly design tool. Ultimately, with the collaboration of ARM and Cadence Design Systems, a full-blown proposal was made to the Open SystemC Initiative (OSCI), under the name PV and PVT. Indeed Programmer view certainly echoes the intent of this new abstraction level, which is to bridge the gap between the embedded software developer and the hardware architect (hardware/software co-development). Certainly, allowing the Algorithm, hardware, software, and functional verification teams to have confidence in the same functional model is saving valuable time by avoiding misunderstandings due to informal or even formal paper-based communication.

2.1 SystemC and Transaction-Level Modeling

SystemC [8, 9] is a language built in standard C++ by extending the language with a set of class libraries created for design and verification. SystemC focuses on the urgency for a system design and verification language that covers hardware and software. SystemC are applied worldwide for doing system-level modeling, abstract analog/mixed-signal modeling, architectural exploration, performance modeling, software development, functional verification, and high-level synthesis [11]. The language is defined by OSCI and ratified as IEEE Std. 1666TM-2011 [8].

TLM standard interfaces for SystemC supplies an important framework required for model exchange within companies and across the IP supply chain [8]. This is intended for architecture analysis, software development, performance analysis, and hardware verification [8, 11]. It explicitly focuses on virtual prototyping in which SystemC models can be exchanged and organized with no difficulty within a system. This is achieved by providing a strong modeling foundation for virtual prototyping. The standard allows optimal reuse of models and modeling effort across different use cases [11].

Use cases have been categorized according to a range of criteria, leading to standard interfaces differentiated by loosely timed (LT) and approximately timed (AT) modeling styles. The extended APIs provide a fundamental, general purpose interoperability layer. A specific payload (generic payload or transaction), to be used in conjunction with these interfaces, helps achieve a higher degree of interoperability when generically modeling memory-mapped bus-based components. Several TLM features boost simulation performance enabling what is called speed interoperability in addition to model interoperability for SystemC virtual platforms. Temporal decoupling allows initiator models, such as instruction set simulators, to run ahead of the SystemC kernel and synchronize only periodically to significantly reduce the required number of costly context switches. The direct memory interface allows interconnect models to be bypassed, facilitating high-speed access to modeled memory. A dedicated transaction debug interface ensures that debugging is an integral part of a system model while enabling debug activity without interference with the system simulation [11]. Figure 1 shows the TLM uses cases with the coding styles, abstractions, and mechanisms.

Different suppliers have adopted TLM broadly due to its interoperability of transaction-level models without compromise to simulation speed. It provides an essential ESL framework for architecture analysis, software development, software performance analysis, and hardware verification. Many companies across the ESL ecosystem already incorporate TLM in their products and solutions [11].

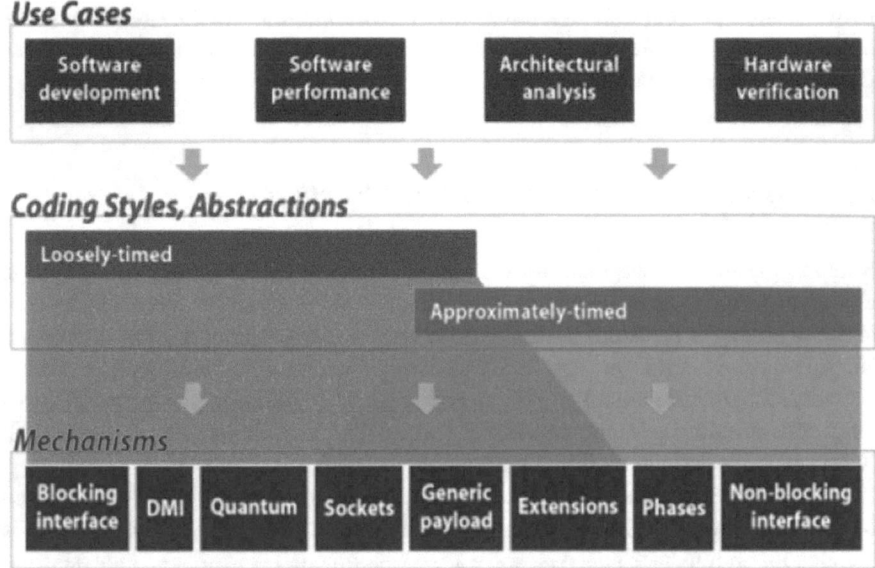

Fig. 1 TLM use cases, coding styles, and mechanisms [11]

2.2 Open Virtual Platform

The open virtual platform [12] is the source of Fast Processor Models and Platforms. The focus of OVP is to accelerate the adoption of the new way to develop embedded software, especially for SoC and MPSoC. OVP has three main components: Open Source Models, OVPsim/Igen/ISS tools, and Modeling APIs. These components makes it very easy to put together advanced multicore heterogeneous or homogeneous platforms with complex memory hierarchies and layers of embedded software, that run at at 100 s of MIPS on standard desktop PCs [12].

The open source models are distributed in several different model categories: pre-compiled object code and source files. Currently, there are processor models of ARC, ARM, MIPS, PowerPC, Renesas, Altera, Xilinx, and OpenRisc families [12]. There are also models of many different types of system components including RAM (Random-access Memory), ROM (Read-only Memory), trap, cache, and bridge. There also peripheral models including DMA (direct memory access), UART (universal asynchronous receiver/transmitter), FIFO (first in, first out), ethernet, and USB (Universal Serial Bus). There are also models of several different pre-built platforms which run operating systems including Linux, Android, MQX, Nucleus, Micrium, and FreeRTOS [12].

The OVPsim simulator is a very fast simulator and is currently released on 32-bit Windows and Linux. It provides the simulation capabilities to run platforms of OVP processor and peripheral models at very fast speeds. The simulator is a just-in-time (JIT) code morphing simulator engine that translates target instructions to x86 host instructions dynamically. It has been specifically designed for the fastest simulation throughput and includes many optimizations enabling simulation of platforms utilizing many homogeneous and heterogeneous processors with many complex memory hierarchies. OVPsim includes very efficient modeling of MMU/TLBs (memory management unit / translation lookaside buffer) and hardware virtualization [12].

OVPsim can be wrapped and called from within other simulation environments and comes as standard with wrappers for C, C++, and SystemC. Another key technology component of OVPsim is that it can encapsulate existing binary models of processors and behavioral models. Thus, the utilization of existing legacy processor models in an OVP simulation is not difficult. OVPsim comes with a GDB (GNU Project Debugger) RSP (Remote Serial Protocol) interface and is easy to use with standard debuggers that support this GDB RSP interface [12]. Moreover, the OVPsim is highly suitable for usage in many educational environment due to its tight research budget characteristic.

3 Virtual Platform Architecture

The virtual platform general specification is to prototype the AES-128 Cryptosystem MPSoC. This includes hardware and software models that are designed and engineered to work together. SystemC, TLM, and OVP technologies are used to build the virtual platform including the hardware system. The architecture of the virtual platform is defined in this section which contains the system partitioning task, the hardware specifications and functionalities, and the software specifications and functionalities. The architecture of the virtual platform (hardware and software) is defined according to the case study presented in Sect. 4.

The cryptosystem takes plaintext as the input data and produces ciphertext as the output data. The size of the data is 128 bit. In the real world, the data will be inserted into the system by another system. The encryption algorithm used within the system is AES. The architecture of the virtual platform is designed to be scalable and configurable to certain extent. The scalability factor is the number of processing elements (PEs). Thus, the user of this virtual platform is able to specify the number of processing elements available within the system during the virtual platform execution (runtime). The configurable factor is the start and end addresses of the memories (RAM and ROM) within the system. The size of the RAM could also be modified on the fly.

3.1 Virtual Platform Specification

The virtual platform is built using TLM-2.0 LT coding style. The functional specifications of the virtual platform are presented in Fig. 2 using UML (Unified Modeling

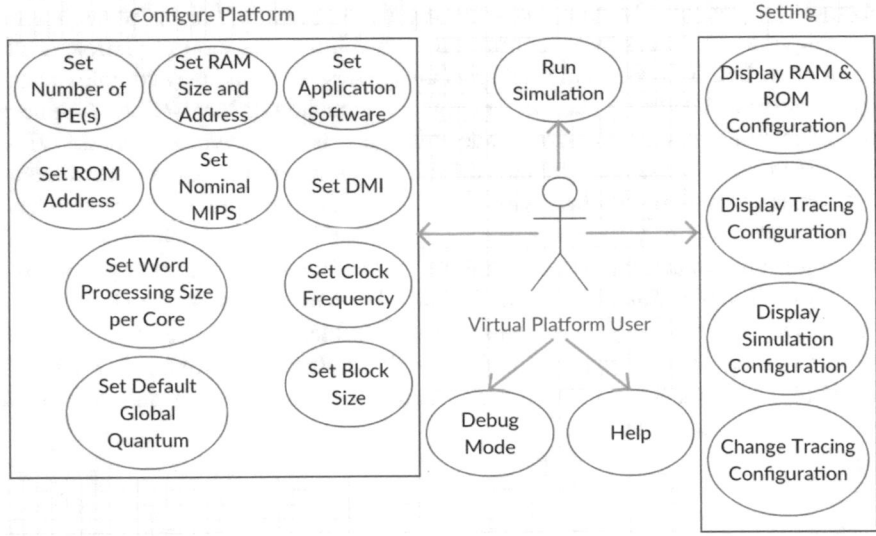

Fig. 2 Virtual platform use case diagram

Language) 2.0 use case diagram for the sake of simplicity and presentation of this paper.

OVPs ARM Cortex-M0 processor models are used in the virtual platform, one in each processing element. Additional features of TLM i.e., temporal decoupling and DMI (direct memory interface) are used within the virtual platform to speed up simulation speed.

The prototyped system is partitioned into hardware and software. The implementation of the encryption is done in the hardware. The software sets the encryption key, and the plain text is from another hardware that is called host device in this paper. The host device is developed as a TLM module and is part of the system-level testbench. The functionalities of the virtual platform is built to provide scalability on the hardware architecture that is the number of the processing element and the configurability on the start and end addresses of the memories.

3.2 Hardware Architecture

The default hardware architecture with four processing elements is displayed in Fig. 3. The processing elements are divided into two categories: central and encryption. The central PE contains a DMA controller module, an AHB2PP (AHB to parallel port) module. The other encryption PEs pack an AES accelerator module each. The interconnect system between all of the PEs is NoC with 2D mesh architecture. The bridge between bus-based interconnect system and the NoC is the AHB2NoC

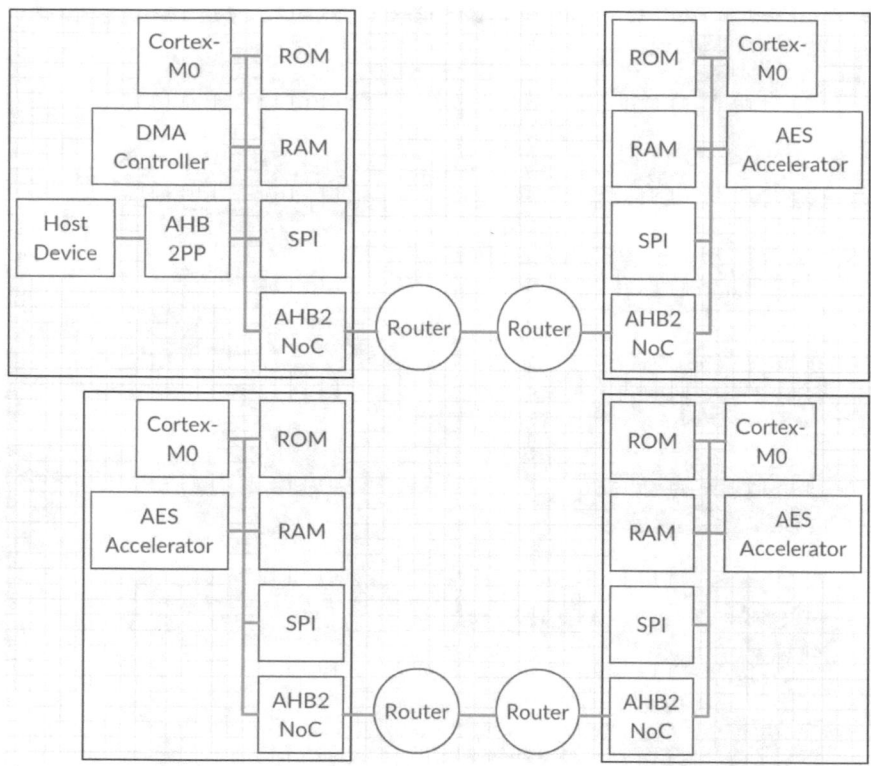

Fig. 3 Virtual platform default hardware architecture

(AHB to NoC) Module. The hardware system is attached to an external module host device as part of the system-level testbench. The Cortex-M0 Processor, ROM, RAM, and bus-based interconnect system are used from the OVP model repository.

The development of the hardware models other than stated previously is divided into three categories: TLM initiator module, TLM target module, and TLM interconnect module. The initiator module initiates transactions and sends it to other modules. The target module accepts transactions and reacts based on the transactions. In timed simulation, the additional delay occurred because of the operation done by the module is added into the transaction. The interconnect module simply resembles the function of an interconnect system, that is to pass the transaction without modifying it, except adding additional timing delay in the timed simulation.

The use of the TLM-2.0 mechanisms increases the modules interoperability and reusability. This has been proven with the integration of OVPs high-level models with the rest of the models in this virtual prototyping platform. A wrapper modules are also developed i.e., the AES accelerator module that wraps the Texas instruments AES model implemented in C programming language [1, 7]. The DMA controller module has four independent channels to support data transfer between processing

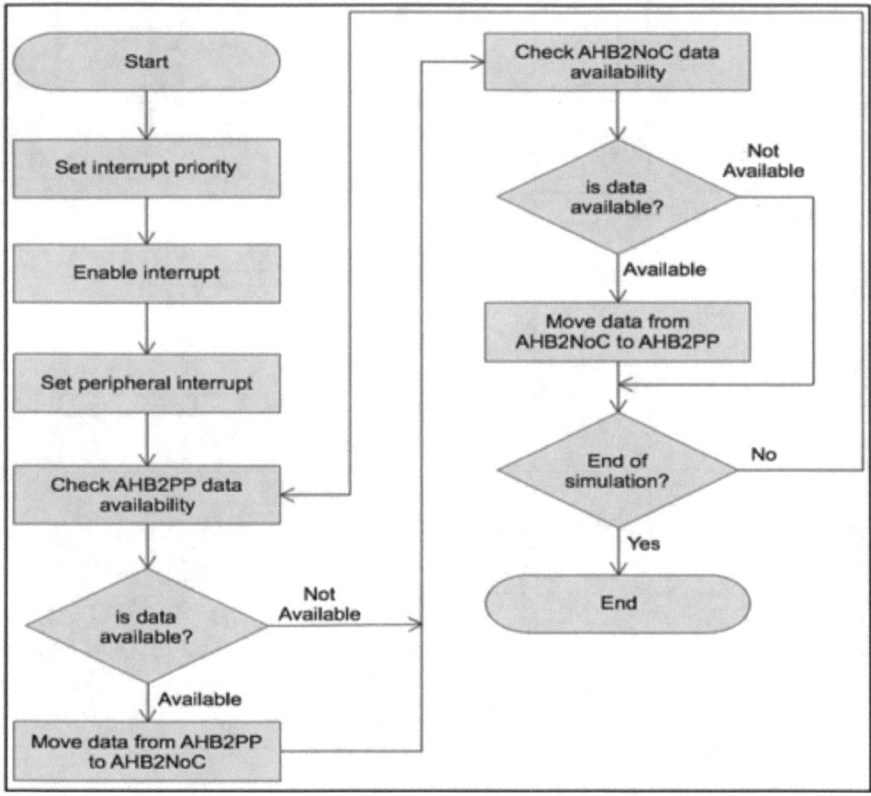

Fig. 4 Central PE firmware flowchart diagram

elements. The implementation of these channels uses *SC_THREAD* to provide the multithreading capability for each of the channels. This way the simulation has several active threads including the processor and the DMA controller modules active channel.

3.3 Software Architecture

The software architecture of the system is highly coupled with the hardware architecture. There is two type of firmware required by the system: central PE firmware (Fig. 4) and encryption PE firmware (Fig. 5). Other than the firmware, the software includes drivers, ISRs (interrupt service routines), and linker scripts.

There are five drivers for each of the hardware peripherals: AHB2PP, AHB2NoC, SPI (Serial Peripheral Interface), DMA controller, and AES accelerator. These drivers provide the interfaces for both central and encryption PE firmware to access

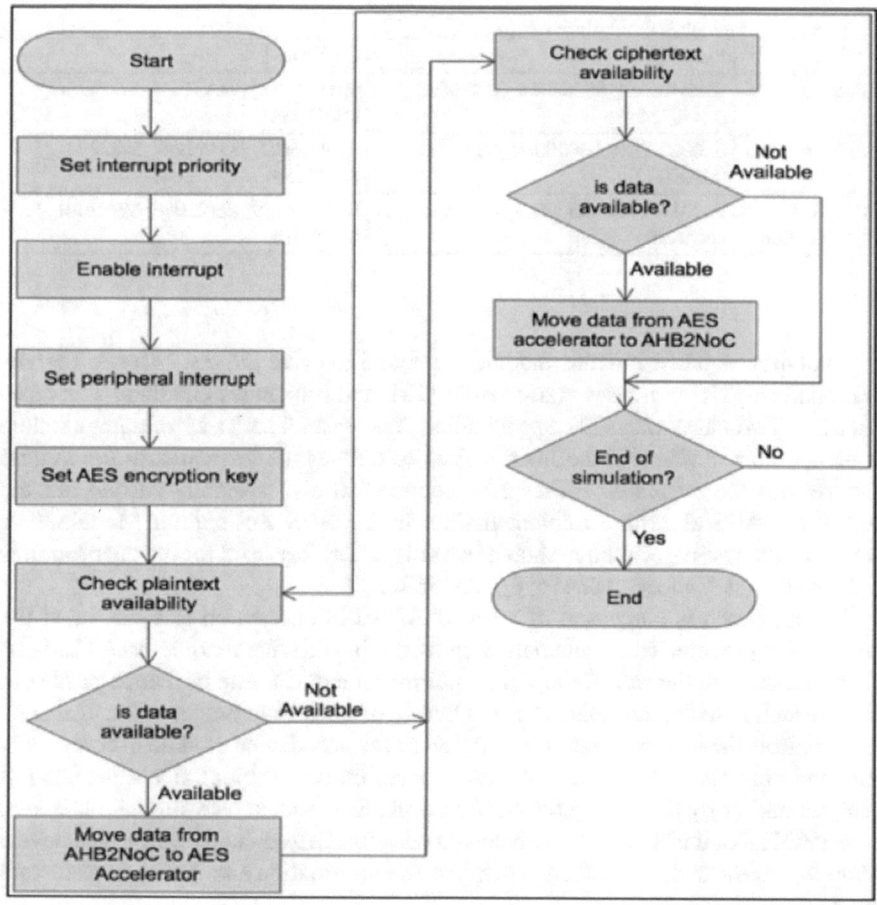

Fig. 5 Encryption PE firmware flowchart diagram

the peripherals internal registers and perform certain tasks. The drivers are cross compiled together with the firmware, ISR, and linker script into an executable binary file using Sourcery CodeBench Lite Edition. The whole encryption process is carried out by multiple processors and programmed by the two firmware. The pure encryption task is done by the encryption PE firmware. Among the encryption PEs, the division of the task is accomplished by coarse-grained parallelism implementation.

3.4 System-Level Testbench

The virtual platform is merely a platform that needs a testbench in order to verify the system-level functionalities of the virtual platform and also the MPSoC that

Table 2 AES-128 encryption test-cases

Test-case	Plaintext	Ciphertext
Gladman	0x3243f6a8_885a308d_313198a2_ e0370734	0x1995a9f6_764a79a4_acddc008_ b805298a
All Zero	0x00000000_00000000_00000000_ 00000000	0x8ffb667f_00df6bdd_5a224bdf_ 8f1d325b
AES Test Specification	0x00112233_44556677_8899aabb_ ccddeeff	0x4c1cce19_def4305c_83bcf0d1_ 355074f8

it prototypes. A TLM initiator module is needed to drive the test-cases to the virtual platform. The test-cases is used from [13], and it includes Gladman Test-case, All Zero Test-case, and AES Specification Test-case. The TLM initiator module is named host device and the main task is to deliver the test-cases to the system and verifies the results once they are produced. It also specially carried out the test of the AES algorithm implementation in the AES Accelerator Module. The details of the test-cases used is shown in Table 2. The key used for the encryption is 0x12345678_12345678_12345678_12345678.

The testbench is composed of a set of AES-128 encryption test-cases and the host device module. The implementation of the host device module uses TLM-2.0 LT to match with the rest of the virtual platforms models. The testbench prints out an information to the terminal at the beginning of the testing process. The testbench then controls the repetition number and step of the host device module to completely and randomly transfers all of the test-cases according to the block size value that has been set earlier by the user in the configure platform menu. The simulation is kept alive untill all of the ciphertext are received back by the testbench (via the host device Module). At the end, the testbench displays the information about the test result and quits.

4   A Case Study: AES-128 Cryptosystem MPSoC

The simulation and testing of the virtual prototyping platform is carried on a virtual machine with Centos Red Hat (64-bit) operating system, with 8 GB of RAM DDR3 and an Intel Core i7 processor (2.2 GHz). In addition, five AES-128 benchmark software developed by the UTAR VLSI Research Centre for the RUMPS401 are used for the simulation and testing. They are: AES-128 Benchmark-16, AES-128 Benchmark-32, AES-128 Benchmark-64, AES-128 Benchmark-128, and AES-128 Benchmark-256. These benchmark software are used to measure the functional correctness of the virtual prototyping platform. The simulation is run using the virtual platforms configurations as shown in Table 3.

In this section, the result of practical implementation of a virtual prototyping platform is described. The virtual prototyping platform provides the abilities for

Table 3 Lotus-G configuration for correctness testing

Parameter	Value
PEs	4
MIPS	100
Quantum	1 μs
DMI	Yes
Clock frequency	0
Temporal decoupling	Yes
Tracing and instrumentation	Off

the MPSoC engineers (both hardware and software engineers) to perform hardware/software co-design and co-verification specifically: early software development, architectural exploration, estimated timing analysis, and functional verification.

4.1 Early Software Development

The virtual prototyping platform can be used to develop new software before the RTL platform or the hardware prototype is available. This is achieved by the development of the AES-128 Benchmark-DMA and AES-128 Benchmark-512. This is one of the key result that is important to note: the ability to design the software concurrently with the hardware. There are three architectures that are used to develop the software: 4 PEs, 8 PEs, and 12 PEs. For each of the architectures, different encryption block size is used as the parameter in developing the software, i.e., The 12 PEs and 256 encryption block size uses 9 Main0.c and 2 Main1.c, where each serves 24 blocks and 20 blocks, respectively.

4.2 Architectural Exploration

The simulation and testing performed shows the use of the virtual prototyping platform for system designers to design, merge, and optimize complex systems meeting design specifications such as functionality and performance. Performance could be investigated together with functional verification and software execution.

The DMA controller module is attached and unattached in various configurations as displayed in Table 4. This is performed to demonstrate the ability of the virtual prototyping platform in enabling architecture exploration. The performance results of different hardware architectures are shown in Table 5 and the parameters of the simulations are displayed in Table 6.

Table 4 Architectural exploration test results

Simulation	DMA Controller Module	Untimed (s)	LT (s)
1st	None	0.13	23.41
2nd	One in Central PE	3.40	11.34
3rd	One in each of the PEs	3.52	13.22

Table 5 Architectural exploration test parameters

Parameter	Value
Number of PEs	4
Encryption block size	256
MIPS	1 μs
Global quantum size	100 μs
Clock frequency	32 MHz (LT)
DMI feature	On
Tracing and instrumentation features	Both off

Table 6 Tested parameters range

Parameter	Range value
Number of PEs	1–12
Encryption block size	16–512
MIPS	1–100
Global quantum size	1 μs–1 s
Clock frequency	16 MHz (LT) or 32 MHz (LT)
DMI feature	On or off
Tracing feature	On or off
Instrumentation feature	On or off

The LT simulations give more timing accurate results compare to the untimed simulations. The timing delay produced by the models during the LT simulations is not 100% precise, and it is just an estimate used for the high-level modeling purpose. Thus, more accurate delay time figures could be used to replace the existing value when available. In Table 4, the fastest LT simulations is achieved in the 2nd simulation, where the architecture has only one DMA Controller Module in the Central PE. Deeper and broader exploration could be performed further i.e., varying the number of the PEs, the interconnect systems, and the use of specific hardware accelerator modules in order to get the best architecture for the targeted application. Exploration on the software side could also be performed i.e., the firmware, driver and ISR, specifically the task division, the parallelism and the algorithm.

```
Info ---------------------------------------------------
Info SIMULATION TIME STATISTICS
Info    Simulated time        :  0.01 seconds
Info    User time             :  3.52 seconds
Info    System time           :  0.00 seconds
Info    Elapsed time          :  3.52 seconds
Info ---------------------------------------------------

OVPsim finished: Wed Jun 15 21:09:54 2016
```

Fig. 6 Simulation time statistics example

4.3 Estimated Timing Analysis

The virtual prototyping platform could give estimated timing to the simulation based on the LT implementation of the TLM models. This enables estimated timing analysis to be done. Both hardware and software engineers could observe the consumed time based on the LT simulation. An example is shown in Fig. 6 where the simulation time statistics contains information i.e., simulated time, user time, system time, and elapsed time. Simulated time is the simulation duration in simulated time (LT). User time is the wall-clock time that is assigned to the simulator. System time is the time used for doing system chores on behalf of the simulator process. Elapsed time matches the wall clock from the start of the simulation until the simulation time statistics line is printed.

4.4 Functional Verification

The virtual prototyping platform serves as a tool to verify the functionality of the hardware concurrently with the software. During the simulation, the hardware and the software are essentially verified together in order to achieve the expected outcomes and to behave correctly according to the system functional specifications. A semantic error caused by the software will cause the hardware to misbehave, hence stops the simulation. An error in the hardware will also cause the simulation to hang, and the error is observable by activating the tracing feature.

The DMA controller RTL simulation was performed using Synopsys VCS with a SystemVerilog HVL (hardware verification language) testbench created following the UVM (Universal Verification Methodology). The result compared well with that obtained using the high-level virtual platform.

5 Conclusion

The virtual prototyping platform was built using TLM in SystemC to enable the MPSoC hardware/software co-design and co-verification. It was demonstrated to work successfully in early software development, architectural exploration, estimated timing analysis, and functional verification, all using the same high-level models. It is also scalable and configurable, yet fast and accurate enough. In future, this technique will be extended to IoT (internet of things) systems with cryptographic security system development in mind.

Acknowledgements My deepest gratitude to my supervisor Mr. Tang Chong Ming, Prof. Lee Sze Wei, and Mr. Ng Mow Song for the support of this research, for the encouragement, enthusiasm, and immense knowledge. This would not have been possible without their guidance and support.

References

1. AES-128 Advanced Encryption Standard | TI.com. http://www.ti.com/tool/aes-128. Accessed 17 Feb 2017
2. Bailey, B., Martin, G.: ESL Models and Their Application. Springer, Boston (2010)
3. Bailey, B., Piziali, A.: ESL Design and Verification. Morgan Kaufmann, Amsterdam (2010)
4. Cota, E., et al.: Reliability, Availability and Serviceability of Networks-on-Chip. Springer, Boston (2012)
5. Engblom, J., Aarno, D.: Software and System Development Using Virtual Platforms: Full-system Simulation With Wind River Simics. Morgan Kaufmann Publishers (2015)
6. Ghenassia, F.: Transaction-Level Modeling with SystemC. Springer, Dordrecht (2005)
7. Hall, J.: C Implementation of Cryptographic Algorithms. Texas Instruments (2013)
8. IEEE Standard for Standard SystemC Language Reference Manual. Institute of Electrical and Electronics Engineers, New York (2012)
9. Leupers, R., Temam, O.: Processor and System-on-Chip Simulation. Springer (2010)
10. SystemC Community. http://accellera.org/community/systemc. Accessed 17 Feb 2017
11. SystemC TLM. http://accellera.org/community/systemc/about-systemc-tlm. Accessed 17 Feb 2017
12. Technology. http://www.ovpworld.org/technology. Accessed 17 Feb 2017
13. Wagner, N.: The Laws of Cryptography: Test Runs of the AES Algorithm. http://www.cs.utsa.edu/~wagner/laws/AEStestRuns.html. Accessed 17 Feb 2017
14. Wolf, W., Jerraya, A.: Multiprocessor Systems on Chips. Elsevier, Amsterdam (2005)

A Data-Mining Model for Predicting Low Birth Weight with a High AUC

Uzapi Hange, Rajalakshmi Selvaraj, Malatsi Galani
and Keletso Letsholo

Abstract Birth weight is a significant determinant of a newborn's probability of survival. Data-mining models are receiving considerable attention for identifying low birth weight risk factors. However, prediction of actual birth weight values based on the identified risk factors, which can play a significant role in the identification of mothers at the risk of delivering low birth weight infants, remains unsolved. This paper presents a study of data-mining models that predict the actual birth weight, with particular emphasis on achieving a higher area under the receiver operating characteristic (AUC). The prediction is based on birth data from the North Carolina State Center for Health Statistics of 2006. The steps followed to extract meaningful patterns from the data were data selection, handling missing values, handling imbalanced data, model building, feature selection, and model evaluation. Decision trees were used for classifying birth weight and tested on the actual imbalanced dataset and the balanced dataset using synthetic minority oversampling technique (SMOTE). The results highlighted that models built with balanced datasets using the SMOTE algorithm produce a relatively higher AUC compared to models built with imbalanced datasets. The J48 model built with balanced data outperformed REPTree and Random tree with an AUC of 90.3%, and thus it was selected as the best model. In conclusion, the feasibility of using J48 in birth weight prediction would offer the possibility to reduce obstetric-related complications and thus improving the overall obstetric health care.

U. Hange (✉) · R. Selvaraj · M. Galani · K. Letsholo
Department of Computer Science & Information Systems, Botswana International University
of Science and Technology, Palapye, Botswana
e-mail: uzapi.hange@studentmail.biust.ac.bw

R. Selvaraj
e-mail: selvarajr@studentmail.biust.ac.bw

M. Galani
e-mail: galanim@biust.ac.bw

K. Letsholo
e-mail: letsholok@biust.ac.bw

© Springer International Publishing AG 2018
R. Lee (ed.), *Computer and Information Science*, Studies in Computational
Intelligence 719, DOI 10.1007/978-3-319-60170-0_8

Keywords Birth weight · Low birth weight · Data-mining · SMOTE · Imbalanced dataset

1 Introduction

Low birth weight (LBW) is often classified as follows: very low (<1, 500 g), low (<2,500 g), or normal (at least 2,500 g) weight at birth [1]. Evidence shows that LBW newborns are at a higher risk of perinatal and infant mortality and other health complications such as mental retardation, and respiratory distress syndrome [1–5].

According to the World Health Organization (WHO), about 15.5% of births worldwide are LBW. Reducing cases of low birth weight by at least one-third is among the seven aims for the current decade of the "A world fit for Children" programme of the United Nations Children's Fund [2]. As a result, there is an intensifying demand for robust and noninvasive methods of estimating birth weight and predicting newborns at a higher risk of being born with LBW [6]. Accurate prediction of LBW permits early obstetric interventions [6–8]. Thus, of recent, data mining has emerged as one of the most useful solutions to predict LBW. Data mining has been defined as a field that unearths new and useful trends from large historical datasets. It achieves its goal by mining patterns from huge datasets and consequently builds models that could later perform predictive or descriptive tasks [9].

There are few studies focused on LBW prediction using data mining [4, 6], and these few studies are largely based on descriptive data mining. Thus, previous studies have paid a great deal of attention to capturing important genetic, demographic, and obstetric factors associated with low birth weight prevalence. The identified factors can be profiled and used to make predictions on birth weight outcomes [10]. However, the applicability of data mining to estimate the actual birth weight based on the identified risk factors has not been fully studied.

As such, it is necessary to build a model that predicts future birth weight outcomes based on the identified factors rather than exhaustively listing risk factors for LBW. Therefore, the study proposes a data-mining model that estimates birth weight values for unborn babies, with a relatively higher AUC. Model selection is accomplished by comparing performance measures of three decision tree models; C4.5, REPTree, and Random Tree.

In addition, a special technique, SMOTE, was applied to advance the AUC of decision trees, since the AUC has greater value than the classifier accuracy in clinical settings [10, 11]. This was achieved by increasing the cases of the under-represented birth weight classes to handle imbalanced clinical data.

This research attempts to answer the following question: Can decision trees built from a dataset with artificially increased cases of the minority class yield a higher AUC than the ones built from raw imbalanced data?

2 Literature Review

2.1 Data-Mining Classification

Classification is a supervised leaning method, whose principal objective is to build models based on known data and predict the category for new data, using the built model [12–14]. In classification, dividing a supplied dataset into training and test sets develops models. The training set is run through one or numerous classification algorithms, and the classifier models are subsequently built. The test set is then used to evaluate the performance of the models. Lastly, the classifier model together with various evaluation metrics is presented, demonstrating its capability to correctly classify new cases [12]. Decision tree learning algorithms are among the classification methods that frequently cited in the literature and are discussed in the subsequent section.

2.2 Decision Tree Classification

According to Gupta et al. [14], decision trees are flowchart-like trees where each internal node depicts a test on an attribute, each branch represents an outcome of the test, and leaf nodes represent classes.

In this study, decision trees were used for classification and prediction due to the following reasons; they have been reported to perform relatively better than other classification algorithms, not only in their construction but also in accuracy as well. Furthermore, they can deal with high-dimensional data without compromising their performance. Another notable benefit of decision trees is their capability to display model results in comprehensible formats [15].

2.3 Related Work

In the past years, researchers have made efforts to predict LBW. Yadav and Lee [3] applied statistical analysis to LBW data to build a LBW predictive model. This cross-sectional study was conducted at Seremban General Hospital in Malaysia between May 2007 and March 2008. The total number of records was 666, 573 normal and 93 LBW births. Logistic regression was used to develop a LBW predictive model. The model was used to forecast the factors associated with LBW. The LBW model yielded a sensitivity rate of 80% and specificity of 75% using ROC curve.

A study by Marshall et al. [16] developed a model for very LBW neonatal mortality prediction, using historical birth data in 16 neonatal intensive care units from Argentina, Chile, Paraguay, Peru, and Uruguay. Infants weighing 500–1500 g from the October 01, 2000 to the May 30, 2003 were included in this study. The model was developed using 1, 801 instances, while 450 instances served as a test sample. This model was compared with two existing scores by using the AUC, and it was highly predictive for in-hospital mortality with 85% AUC.

Senthilkumar and Paulraj [4] applied a wide range of data-mining algorithms to predict LBW babies. The algorithms were Naïve Bayes', neural network, random forest, classification tree, SVM, and logistic regression. A dataset from Baystate Medical Center, Springfield, Massachusetts, was used. The classification tree yielded a higher prediction accuracy of 89.95% when using cross-validation method on training data. Random forest gave a lower overall prediction accuracy 70.9% than other algorithms. The results of the study were compared with that of logistic regression in other works, and logistic regression's prediction accuracy was outperformed by the data-mining techniques employed in this study.

Desalegn [6] introduced J48 and PART rule induction algorithm (with pruning) in predicting LBW using Ethiopia demographic and health survey data. Experiments were conducted on a dataset containing 9, 861 records dataset. J48 decision tree classifiers using pruned technique achieved an accuracy rate of 94.7%. PART rule induction algorithm gave an accuracy rate of 94.35%.

The examined studies yielded substantial results. However, what remains unsolved is the use of data-mining techniques for predicting birth weight based on LBW risk factors. Hence, this study attempts to bridge this research gap.

3   Methodology

3.1   Data Selection

The research utilized a publicly available dataset containing 10, 000 births recorded at the North Carolina State Centre for Health Statistics in 2006. The target variable was the birth weight group (grams), and it had 10 classes: 0 (500 or less), 1 (501–1000), 2 (1001–1500), 3 (1501–2000), 4 (2001–2500), 5 (2501–3000), 6 (3001–3500), 7 (3501–4000), 8 (4001–4500), and 9 (4501 or more). Initially, the dataset contained 131 predictor variables. In this section, relevant variables for model building were selected. The data file description was used to gain understanding of the variables. Anomalies and redundant variables were eliminated. The final number of attributes for initial modeling was 90 and was exported to Waikato Environment for Knowledge Analysis (WEKA) workbench for preprocessing.

3.2   Handling Missing Values

Missing data can significantly impact models' performance. Our dataset had a considerable amount of missing values, and the study employed a feature named "ReplaceMissingValues" in WEKA to automatically replace missing values. This feature replaces continuous missing values with the mean and nominal missing values with the mode. Attributes with a large number of missing values were discarded from analysis. Age and education of the father attributes accounted for approximately 18% of missing values, and, therefore, they were deleted.

3.3 Handling Imbalanced Data

The data imbalance problem is realized when the number of instances in one class is much smaller (minority class) than the instances in another class (majority class) [17–19]. Building predictive models based on imbalanced datasets and multiple classes can be overwhelming and countering imbalance measures are necessary [11]. WEKA visualizations revealed that the dataset was imbalanced. Class 6 (3001–3500 g) was the most frequent with 3, 769 instances. It was followed by class 7 (3501–4000 g) with 2, 660 instances, then class 5 (2501–3000 g) with 1, 843 instances, and then Classes 8 (4001–4500 g) and 4 (2001–2500 g) with 720 and 531 instances, respectively. The least frequent classes were classes 0 (500 or less grams), 1 (501–1000 g), 2 (1001–1500 g), 9 (4501 or more), and 3 (1501–2000) with 32, 80, 75, 112, and 178 instances, respectively. The frequency per class is shown in Fig. 1.

Our objective was to bridge the gap between the minority and majority class distributions using SMOTE. This technique was applied to the dataset to create synthetic instances for the minority classes (0, 1, 2, 3, 4, 5, 8, and 9).

Depending upon the amount of over-sampling required, neighbors from the k nearest neighbors are randomly chosen [17]. For instance, if the over-sampling needed is 200%, only two neighbors from the five nearest neighbors are chosen and one sample is generated in the direction of each [19]. Our approach used ten nearest neighbors. The number of synthetic cases generated in each iteration is shown in Fig. 2.

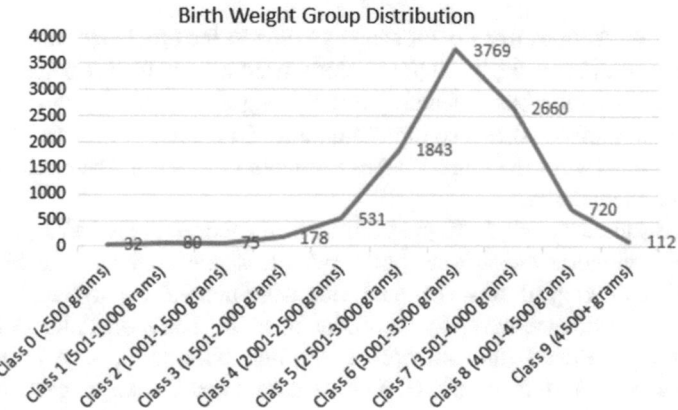

Fig. 1 Birth weight frequencies per class

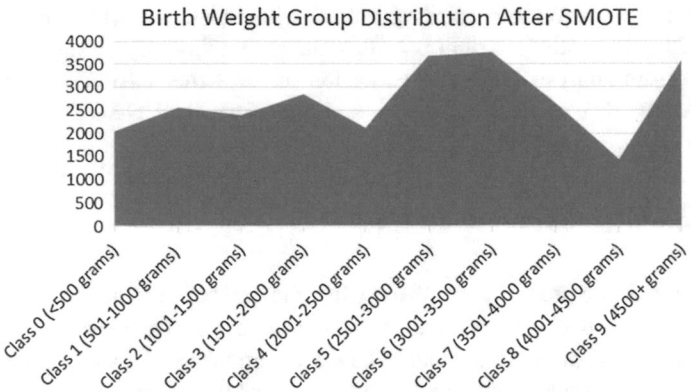

Fig. 2 Final distribution per class, after SMOTE 600%

Initially Class 0 had 32 cases, and after SMOTE 600% it had 2, 048 cases. Class 1 was comprised of 80 cases before SMOTE and 560 cases after SMOTE 500%. Class 2 entailed 75 cases, and after SMOTE 500% it had 2, 400 cases. Class 3 contained 178 cases prior to SMOTE and 2, 848 after SMOTE 400%. Class 4 originally had 531 cases and 2, 124 after SMOTE 200%. The initial class distribution for Class 5 was 1, 843 cases and 3, 686 after SMOTE 100%. The distributions for Class 6 and Class 7 were maintained since no SMOTE was implemented on the two classes. Class 6 had 3, 769 while Class 9 had 2, 660 cases. Class 8 had 720 cases and a final class distribution of 2, 880 after SMOTE 200%. Finally Class 9 had 112 cases and a total of 3, 584 cases after SMOTE 500%. Figure 2 shows the final distribution in each class after SMOTE 600%.

3.4 Feature Selection

By eliminating features that do not add any value to the efficiency of the algorithm, performance is enhanced [20, 21]. The study used a filter feature selection method named the correlation-based feature selection algorithm (CFS), coupled with the best fit search method to reduce dimensionality. CFS searches for features that are highly correlated with the target classes yet have minimal inter-correlation among the features themselves.

Figure 3 indicates that Total preg (number of total pregnancies) and Meconium (Meconium, moderate/heavy) attributes were least correlated to the birth weight group with a ranking of 10% each and thus were manually eliminated.

County of residence of mother, Gender of child, Race of child, Education of mother (years), Bdead (the number of children born alive now Dead), Terms (number of other terminations), Gest Age (completed weeks of gestation (calculated)), Numchild (number of living children), Avecigs (average number of

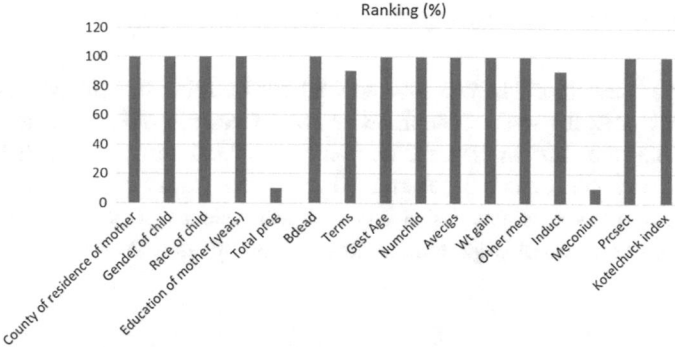

Fig. 3 Rankings by CFS algorithm and best search strategy

cigarettes used daily), Wt gain (weight gained), Other med (other medical conditions), induct (Induction of labor), Prcsect (repeat cesarean section), and Kotelchuck index were all highly correlated to the birth weight group with a score of 100% and therefore were retained for model building.

3.5 Model Building Using J48

The J48/C4.5 decision tree has been extensively applied to medical problems. C4.5 [22, 23] is a successor of ID3, developed by Quinlan Ross (1993). Comparative to earlier versions of decision trees, C4.5 algorithm takes both continuous and categorical attributes as training input. It divides continuous data into discrete intervals; this is widely known as discretization. The data is arranged at every node in order to determine the best splitting feature using gain ratio impurity technique. The construction of a J48 classifier in predicting the birth weight group was done using default parameter settings in the Weka explorer.

3.6 Model Building Using REPTree

The other classification technique applied in this study is the REPTree (reduced-error pruning tree) algorithm. REPTree is a fast classifier that builds a classification/regression tree via information gain/variance. It prunes its trees by means of reduced-error pruning (with back-fitting). REPTree algorithm only sorts the instances for numeric variables once. The algorithm also utilized the default values provided by WEKA.

3.7   Model Building Using Random Tree

The Random tree algorithm was also applied for building the classification model based on the default parameter settings. In recent years, the Random tree learning algorithm has received immense attention in medical research. With k random variables at each node, a tree is constructed at random from a wide range of possible trees. Each tree has an equal probability of being selected. Random trees are very efficient, and the fusion of large Random trees normally leads to accurate models [24].

3.8   Model Evaluation

As novel models are being developed, performance measures are essential to guide research in promising directions. Experimental results based on the test set were evaluated in terms of the details provided by the ROC curve. The sensitivity and specificity were also recorded.

Sensitivity is also known as the true positive rate (TPR) or recall; this is the ratio of the number of positive instances classified over all the positive instances [25]. Sensitivity = True Positive (TP)/True Positive (TP) + False Negative (FN)).

Specificity is used for the purpose of measuring the proportion of negative cases that were correctly classified as negative, which is 1—FP (False positive), [11, 25] or can be denoted as follows: Specificity = True Negative (TN)/True Negative (TN) + False Positive (FP).

The ROC curve [26] is a two-dimensional graph representing the ratio of false positive and true positive rate. On an ROC curve, the X-axis represents the percentage of the FP (1 − specificity) = FP/(TN + FP) and the Y-axis represents the TP (sensitivity) = TP/(TP + FN). The AUC is a standard performance metric for a ROC curve. It takes any value between [0, 1]. A random classifier yields an AUC of ~0.5. A larger AUC represents a better model [26].

4   Results and Discussion

4.1   Experimental Design

Predictive model building is an iterative procedure, and, therefore, it is crucial to perform multiple experiments with different classifiers in order to select the best model for solving the problem at hand [11]. The experiments were conducted by using a meta-classifier in conjunction with J48. REPTree and Random tree and discretization. The performance of the algorithms was evaluated on both the unbalanced and balanced datasets.

Experimentations were carried out for each algorithm, and percentage split validation was used to estimate the performance of each algorithm. Out of all the instances, 80% was utilized for training and the remaining 20% made up the test set. Overall sensitivity, specificity, and AUC were used to evaluate and compare the performance of the models. The AUC serves as an indicator of the overall performance of the algorithm [27]. Therefore, models with the highest AUC were deemed as the best.

The experiments were generally carried out in two phases;

1. The first phase was performed on the actual imbalanced data.
2. The second phase made use of instances generated by SMOTE algorithm (balanced data) and the best selected attributes.

4.2 Experimentation One (Model Building Using Imbalanced Data)

This experimentation made use of the original dataset (imbalanced data) with 10, 000 instances and 90 features. The original dataset was analyzed to determine whether decision trees built from a dataset with artificially increased cases of the minority class possibly yield a higher AUC than the ones built from raw imbalanced data. A total of 8, 000 instances were used for training, and all the three classifiers were used. Performance evaluation was done on the remaining 2, 000 instances. A summary of the obtained output from all the three different experiments based on the test data is represented in Table 1.

As illustrated in Table 1, the models were almost incapable of classifying the positive instances as positive when applied to the imbalanced dataset. This is signified by the sensitivity. REPTree gave the highest sensitivity of 39.2%, J48 came after REPTree with 36.8%, and Random tree was the least sensitive to positive instances with a sensitivity rate of 28.6%. In contrast, the specificity rates for all the models were relatively higher than their sensitivity rates. Thus, the models can better recognize the classes that are not low birth weight compared to the low birth weight classes. The specificity for the models varied with a relatively small margin. J48 had the best specificity of 74.9%, followed Random tree with 74.7%, whereas REPTree had 73.4%.

Table 1 Performance output of models when applied to an imbalanced dataset

Performance measure	J48	REPTree	Random tree
Sensitivity/Recall (%)	36.8	**39.2**	28.6
FP rate	0.251	0.266	0.253
Specificity (1 − FP) * 100	74.9	**73.4**	74.7
AUC (%)	59.9	**62.9**	51.8

For this experimentation, the REPTree model predicts birth weight with a comparatively higher AUC. The REPTree model obtained the maximum AUC score of 62.9%, followed by J48 model with 59.9%. The lowest AUC was attained by Random tree (51.8%). REPTree also gave the highest sensitivity of 39.2%, and J48 came second with 36.8%, whereas Random tree had the worst sensitivity (28.6%). Thus, the models' ability to distinguish all the classes was inadequate.

To sum up, the results show that REPTree produced the best results. However, all it has been observed that all the models were better at predicting the majority classes than the classes that had the least number of instances. Thus, our investigation reveals that class imbalance degrades the performance of decision trees in predicting the birth weight. The results of this experimentation would not be applicable in predicting birth weight in clinical settings, thereby heightening the demand for further experimentations. This experimentation was compared to the next experimentation that was performed with the SMOTE algorithm.

4.3   Experimentation Two (Model Building Using SMOTE)

This experiment made use of 28, 559 instances generated by SMOTE algorithm. Out of these instances, 80% (22, 848) were used for training and the remaining 20% (5, 711) for performance evaluation. In this section, we present the results of the algorithms when implemented on data that was modified with SMOTE algorithm, and with the 14 best selected attributes.

Comparative to experimentation one, Table 2 depicts that there was a substantial enhancement in the sensitivity of all the models. Thus, the ability of the model to classify the low birth weight classes was intensified. This could be more interesting to medical staff, as the aim of the model is to correctly classify low birth weight classes. The J48 model scored the highest sensitivity of 66.3%, followed by Random Tree with 65.0% and lastly REPTree with 61.7%.

The specificity rates for each model were still higher than the sensitivity rates; however, the margin between the two performance measures was significantly lower compared to experimentation one. This indicates the models' recognition of the positive class as opposed to the negative class increased in this experiment.

The Random tree outperformed the REPTree and J48 with the best specificity of 95.5%. J48 came second with 95.4%, and REPTree came last with 94.9%.

Table 2 Performance output of models built with SMOTE

Performance measure	J48	REPTree	Random tree
Sensitivity/Recall (%)	**66.3**	61.7	65.0
FP rate	0.046	0.051	0.045
Specificity (1 − FP) * 100	95.4	94.9	**95.5**
AUC (%)	**90.3**	88.4	85.3

Similarly, we realized a significant increase in the AUC of all the three models. J48 had 59.9% in experiment one and 90.3% in experimentation two. This was also the highest AUC obtained in this experiment. REPTree had the highest AUC of 62.9%; however, in experiment two, it came second with 88.4%. Likewise, Random tree had an increase from 51.8 to 85.3%. This was the least obtained AUC in experiment two.

In conclusion, all the models from experimentation two were capable of differentiating between the LBW groups and the normal birth weight groups. Therefore, it would be viable to use the models for future predictions. However, the results suggest that the J48 model built using SMOTE would be more of value to medical practitioners.

Comparative to related work in the literature, the SMOTE optimized J48 model outperformed the results of a study by Marshall et al. [16]. The J48 model produced an AUC of 90.3% while their model had 85% AUC.

In terms of the sensitivity, a model by Yadav and Lee [3] yielded a sensitivity of 80%, whereas our J48 model produced a lower sensitivity of 66.3%. The poor results can be partially imputed to varying approaches; our model predicts the actual birth weight values, which is novel. On the other hand, the study by Yadav and Lee [3], lists low birth weight risk factors. In addition, the two models employed different datasets, and, therefore, it would be rational to produce different results. The other related works evaluated performance based on the accuracy and thus were excluded in this discussion.

5 Conclusions and Future Work

The study illustrated the utility of predictive data mining in making early obstetric interventions. In this study, we presented models based on decision tree modeling in order to predict the birth weight of baby so as to identify mothers at a risk of delivering low birth weight babies. The necessary steps required to achieve this goal were fully described, and experiments with different decision trees were conducted and compared.

The results indicated that the J48 model yielded the best results in terms of the AUC. Since the medical field demands prediction with a higher AUC, it is concluded that the J48 models built using SMOTE and the best selected features would be practically useful for future birth weight prediction.

The study discovered that birth weight prediction models optimized by artificially increasing the minority class instances using SMOTE yield significantly better results relative to models built from actual raw imbalanced data. The optimized birth weight prediction models improved the sensitivity, specificity, and the AUC. Thus, the research question of this study was answered.

Although encouraging results were obtained, it was noted that further research efforts need to be conducted to fully utilize the potential of data-mining technology in improving the obstetric health care. In particular, the following areas were identified as deserving further research work;

It has been observed that the performance of classification algorithms differs with datasets. Therefore, additional datasets must be considered to assess the applicability of data mining in predicting birth weight based on other low birth weight risk factors.

Furthermore, it is essential to conduct new experiments based on other data-mining methods such as artificial neural networks and support vector machines since they have also proved to be important data-mining methods in the health care.

References

1. Reichman, N.E.: Low birth weight and school readiness. Future Child. **15**(1), 91–116 (2005)
2. United Nations Children's Fund and World Health Organization: Low birth weight, country regional and global estimates (2004)
3. Yadav, H., Lee, N.: Maternal factors in predicting low birth weight babies. Med. J. Malays. **68** (1), 44–47 (2013)
4. Senthilkumar, D., Paulraj, S.: Prediction of low birth weight infants and its risk factors using data mining techniques. In: Proceedings of the 2015 International Conference on Industrial Engineering and Operations Management, pp. 186–194 (2015)
5. Shittu, A.S., Kuti, O., Orji, E.O., Makinde, N.O., Ogunniyi, S.O., Ayoola, O.O., Sule, S.S.: Clinical versus sonographic estimation of foetal weight in Southwest Nigeria. J Heal. Popul. Nutr. **25**(1), 14–23 (2007)
6. Desalegn, B.: Predicting Low Birth Weight Using Data Mining Techniques on Ethiopia Demographic and Health Survey Data Sets. Addis Ababa University (2011)
7. Salomon, L.J., Bernard, J.P., Ville, Y.: Estimation of fetal weight: reference range at 20–36 weeks' gestation and comparison with actual birth-weight reference range. Ultrasound Obs. Gynecol. **29**, 550–555 (2007)
8. Torloni, M.R., Sass, N., Sato, J.L., Renzi, A.C.P., Fukuyama, M., de Lucca, P.R.: Clinical formulas, mother' s opinion and ultrasound in predicting birth weight. Sao Paulo Med. J. **126** (3), 145–149 (2008)
9. Soni, J., Ansari, U., Sharma, D., Soni, S.: Predictive data mining for medical diagnosis: an overview of heart disease prediction. Int. J. Comput. Appl. **17**(8), 43–48 (2011)
10. Catley, C., Frize, M., Walker, C.R., Petriu, D.C.: Predicting high-risk preterm birth using artificial neural networks. IEEE Trans. Inf Technol. Biomed. **10**(3), 540–549 (2006)
11. Tefera, M.: Application of Data Mining to Predict Urinary Fistula Surgical Repair Outcome. Addis Ababa University (2012)
12. Kaur, H., Wasan, S.K.: Empirical study on applications of data mining techniques in healthcare. J. Comput. Sci. **2**(2), 194–200 (2006)
13. Jeyarani, D.S., Anushya, G., Rajeswari, R.R., Pethalakshmi, A.: A comparative study of decision tree and Naive Bayesian classifiers on medical datasets. Int. J. Comput. Appl. 5–7 (2013)
14. Gupta, S., Kumar, D., Sharma, A.: Data mining classification techniques applied for breast cancer diagnosis and prognosis. Indian J. Comput. Sci. Eng. **2**(2), 188–195 (2011)
15. Yahia, M.E., El-taher, M.E.: A new approach for evaluation of data mining techniques. Int. J. Comput. Sci. Inf. Issues **7**(5), 181–186 (2010)

16. Marshall, G., Tapia, J.L., Ivonne, D., Grandi, C., Barros, C., Alegria, A., Standen, J., Panizza, R., Bancalari, A., Lacarruba, J., Fabres, J.: A new score for predicting neonatal very low birth weight mortality risk in the NEOCOSUR south American network. J. Perinatol. **25**, 577–582 (2005)
17. Chawla, N.V., Bowyer, K.W., Hall, L.O., Kegelmeyer, W.P.: SMOTE: synthetic minority over-sampling technique. J. Artif. Intell. Res. **16**, 321–357 (2002)
18. Khalilia, M., Chakraborty, S., Popescu, M.: Predicting disease risks from highly imbalanced data using random forest. BMC Med. Inform. Decis. Mak. **11**(51), 1–13 (2011)
19. Taft, L.M., Evans, R.S., Shyu, C.R., Egger, M.J., Chawla, N., Mitchell, J.A., Thornton, S.N., Bray, B., Varner, M.: Countering imbalanced datasets to improve adverse drug event predictive models in labor and delivery. J. Biomed. Inform. **42**, 356–364 (2009)
20. Kumar, V., Minz, S.: Feature selection: a literature review. Smart Comput. Rev. **4**(3), 211–229 (2014)
21. Setiono, R.: Feature selection : an ever evolving frontier in data mining. In: JMLR: Workshop and Conference Proceedings, pp. 4–13 (2010)
22. Lakshmi, K.R., Kumar, S.P.: Utilization of data mining techniques for prediction of diabetes disease survivability. Int. J. Sci. Eng. Res. **4**(6), 933–942 (2013)
23. Mazid, M.M., Ali, A.B.M.S., Tickle, K.S.: Improved C4.5 Algorithm for Rule Based Classification
24. Ravichandran, S., Srinivasan, V.B., Ramasamy, C.: Comparative study on decision tree techniques for mobile call detail record. J. Commun. Comput. **9**, 1331–1335 (2012)
25. Cios, K.J., Moore, G.W.: Uniqueness of medical data mining. Artif. Intell. Med. **26**, 1–24 (2002)
26. Huang, J., Ling, C.X.: Using AUC and accuracy in evaluating learning algorithms. IEEE Trans. Knowl. Data Eng. **17**(3), 299–310 (2005)
27. Tanner, L., Schreiber, M., Low, J.G.H., Ong, A., Tolfvenstam, T., Lai, Y.L., Ng, L.C., Leo, Y.S., Puong, L.T., Vasudevan, S.G., Simmons, C.P., Martin, L., Ooi, E.E.: Decision tree algorithms predict the diagnosis and outcome of dengue fever in the early phase of illness. PLoS Negl. Trop. Dis. **2**(3) (2008)

A Formal Approach for Maintaining Forest Topologies in Dynamic Networks

Faten Fakhfakh, Mohamed Tounsi, Mohamed Mosbah,
Dominique Méry and Ahmed Hadj Kacem

Abstract In this paper, we focus on maintaining a forest of spanning trees in dynamic networks. In fact, we propose an approach based on two levels for specifying and proving distributed algorithms in a forest. The first level allows us to control the dynamic structure of the network by triggering a maintenance operation when the forest is altered. To do so, we develop a formal pattern using the Event-B method, based on the refinement technique. The proposed pattern relies on the *dynamicity aware-graph relabeling systems* (*DA-GRS*) which is an existing model for building and maintaining a spanning forest in dynamic networks. It is based on evolving graphs as a powerful model to record the evolution of a network topology. The second level of our approach deals with distributed algorithms which can be applied to spanning trees of the forest. Through an example of a leader election algorithm, we illustrate our pattern. The proof statistics show that our solution can save efforts on specifying as well as proving the correctness of distributed algorithms in a forest topology.

F. Fakhfakh (✉) · M. Tounsi · A.H. Kacem
ReDCAD Laboratory, FSEGS, University of Sfax, B.P. 1088, 3018 Sfax, Tunisia
e-mail: faten.fakhfakh@redcad.org

M. Tounsi
e-mail: mohamed.tounsi@redcad.org

A.H. Kacem
e-mail: ahmed.hadjkacem@fsegs.rnu.tn

M. Mosbah
LaBRI, Bordeaux INP, University of Bordeaux, CNRS UMR 5800,
33405 Talence, France
e-mail: mohamed.mosbah@labri.fr

D. Méry
LORIA Laboratory, University of Lorraine, CNRS UMR 7503, Nancy, France
e-mail: Dominique.Mery@loria.fr

© Springer International Publishing AG 2018   123
R. Lee (ed.), *Computer and Information Science*, Studies in Computational
Intelligence 719, DOI 10.1007/978-3-319-60170-0_9

1 Introduction

With the proliferation of mobile devices and advances in wireless communication technologies, mobile ad hoc networks (MANETs) [19] have drawn the attention of the research community in the last few years. A MANET is a collection of mobile devices, called nodes, such as laptops, smartphones. These nodes are interconnected by wireless links without the aid of any fixed infrastructure or centralized administration. In MANETs, each node acts both as a host and as a router to forward messages for other nodes that are not within the same radio range. They are free to move and form an arbitrary topology. Then, MANETs are characterized as an extremely dynamic system where links between the nodes change over time.

To model dynamic networks, we use the evolving graph model [12] which consists in recording the evolution of the network topology as a discrete sequence of static graphs. The communication between the nodes can be ensured by a distributed algorithm [20]. The latter is designed to run on interconnected autonomous computing entities in order to achieve a common task. To make designing distributed algorithms easier, we model these algorithms with a local computation model and particularly graph relabeling systems [18]. A graph relabeling system is based on a set of relabeling rules which are executed locally. These rules, closely related to mathematical and logical formulas, are able to derive the correctness of distributed algorithms. In this context, one of the most challenging issues in distributed algorithms is to prove their correctness. This task is more difficult in dynamic networks due to highly dynamic behavior and time complexity. Different approaches have been proposed in the literature in order to redefine distributed algorithms in dynamic networks and prove their correctness [4, 6, 8, 13, 16]. However, the major limitation of the studied works is the lack of consensus about their developments and their proofs. Furthermore, proofs which have been presented are done manually.

Some distributed algorithms can be applied only to a particular type of graphs such as tree, ring. In this paper, we deal with algorithms which operate on a tree-based topology such as election and coloration. A tree in a graph is an acyclic and connected subgraph. A set of disjoint trees is called forest. In dynamic networks, according to [9], the network can be partitioned anytime into several connected components. Each one represents a given cluster of nodes that evolves semi-independently. In this case, we can talk about a forest of spanning trees, where a spanning tree is formed in every connected component. Due to the appearance and/or disappearance of communication links, the topology of the network can change. In order to efficiently construct and maintain tree-based topologies, we propose a formal pattern based on the *dynamicity aware-graph relabeling systems* (*DA-GRS*) model [7]. This model is an extension of graph relabeling systems. To specify our pattern, we use a formal method which provides a real help for expressing correctness with respect to safety properties in the design of distributed algorithms. Our proposed approach extends our previous works [10, 11] which do not consider a particular type of graph. It is based on the *correct-by-construction* paradigm [17]. The latter can be supported by a progressive and incremental process controlled by refinement

[3] of models for distributed algorithms. This process allows us to simplify the proofs and to validate the integration of requirements. The Event-B formal method [1] can support this methodological proposal suggesting proof-based guidelines.

The main contributions of this paper consist of two stages. In the first stage, we propose a formal pattern which allows to construct and maintain tree-based topologies in dynamic networks based on the *DA-GRS* model. In the second stage, we illustrate our proposed pattern by an example of a leader election algorithm. This algorithm consists of distinguishing a unique node, the leader, which can act as coordinator to perform some special role in the network. To show the efficiency of our solution in term of proofs reduction, we present the proof statistics comparing the development of this algorithm with and without using the pattern. Our approach can guide the user to specify other algorithms in dynamic networks. So, we can reduce efforts of proofs and specification.

The remainder of this paper is structured as follows: Sect. 2 presents preliminary notions of the evolving graph model and Event-B formal method. In Sect. 3, we introduce an informal description of the proposed pattern. Section 4 details its formal specification. In Sect. 5, we present a case study which illustrates the efficiency of our solution.

2   Preliminaries

2.1   Evolving Graph Model

The evolving graph model, proposed in [12], aims to represent a formal abstraction of dynamic networks, through the formalization of a time domain in graphs. In this model, a dynamic graph can be decomposed as a discrete sequence of static graphs. Each static graph represents a snapshot of the dynamic network at a given time. As an example, we consider the four snapshots taken at different time intervals of a MANET, as depicted in Fig. 1. Formally, let $S_\mathbb{T} = t_0, t_1, \ldots, t_n$ be an ordered sequence of dates used to capture the static graphs. These dates correspond to every time step in a discrete-time system ($\mathbb{T} \subseteq \mathbb{N}$). Except for t_0 and t_n, each t_i corresponds to one or more topological events that modify the network. Each edge is labeled with the dates of its presence. Let $S_G = G_0, G_1, \ldots, G_{n-1}$ be the sequence of undirected static graphs. Each G_i represents the network topology during the period $[t_i, t_{i+1})$ in the evolving graph g. Finally, let G be the union of all G_i in S_G, called the underlying graph of g (see Fig. 1).

The edges are labeled with the date of their presence. For example, the presence of the edge "*ac*" in Fig. 1 at the dates "0" and "1" is represented in Fig. 2 by an edge "*ac*" labeled "0, 1." Then, the triplet $g = (G, S_G, S_\mathbb{T})$ is the corresponding evolving graph.

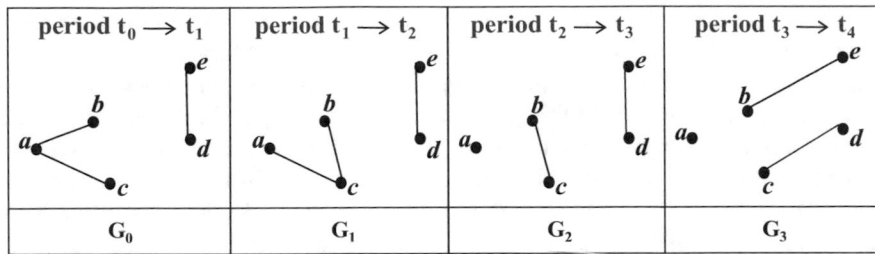

period $t_0 \rightarrow t_1$	period $t_1 \rightarrow t_2$	period $t_2 \rightarrow t_3$	period $t_3 \rightarrow t_4$
G_0	G_1	G_2	G_3

Fig. 1 Successive snapshots of a MANET evolution over time

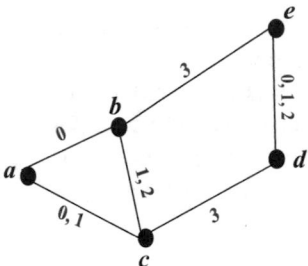

Fig. 2 The evolving graph corresponding to the MANET in Fig. 1

2.2 Event-B Modeling Language

The Event-B modeling language [1] defines mathematical structures into contexts and formal model of system into machines. The modeling process starts by identifying the domain of the problem expressed by means of contexts. A context specifies the static part of a model and may contain *carrier sets*, *constants*, *axioms*, and *theorems* that can be derived from the axioms of a context. An Event-B machine describes a reactive system. It may contain *variables*, *invariants*, *theorems*, and *events*. Variables define the state of a machine. They are constrained by invariants. Theorems are properties derivable from the invariants. Possible state changes are described by events. An invariant is defined to be a predicate preserved by each event. As for an event, it is decomposed into guards that specify under which circumstances it might occur and some generalized substitutions called *actions*. Machines and contexts relationship is defined as follows: A machine M may see a context C, this means that all carrier sets and constants defined in C can be used in M. A machine M' can be built and asserted to be a refinement of the machine M. M' is called a refinement or a concrete version of the machine M. Likewise, a context C' can extend the context C, this means that all properties defined in C' are added to C.

The concept of refinement is the main feature of Event-B. The refinement of a machine allows to enrich it in a *step-by-step* fashion. It is the foundation of the *correct-by-construction* approach [17]. It is also used to transform an abstract model into a more concrete version by modifying the state definition. In fact, new variables and events can be introduced. Furthermore, abstract events can be refined to more concrete ones. The relation between the variables in the concrete and abstract models is given by a *gluing invariant*.

An Event-B specification is considered as correct only if each machine, as well as the process of refinement, is proved by adequate *proof obligations* (*POs*); that is, events preserve the invariant(s) and each event is feasible. POs are generated by the RODIN tool [2], which provides an environment for developing *correct-by-construction* models for software-based systems. They can be discharged either automatically by an integrated proof tool or through interactive proof steps.

3 Informal Pattern Presentation

In software engineering, the idea of design patterns [15] is to have a general and reusable solution to commonly occurring problems. In general, a design pattern is not necessarily a finished product, but rather a template on how to solve a problem which can be used in many different situations. In this section, we propose a formal pattern for specifying and proving the correctness of distributed algorithms in dynamic networks. It can be applied only to algorithms which operate on a tree-based topology. The proposed pattern defines the different topological changes in a dynamic network and the manner of time evolution. Let $g = (G, S_G, S_\mathbb{T})$ be an evolving graph. Every static graph, $G_i \in S_G$, corresponds to the network topology during the interval of time $[t_i, t_{i+1})$ where t_i represents the date when one or several topological events occur in the system. In this paper, we take into consideration only the appearance and disappearance of edges in the network like the existing works in this context [4, 6, 8, 16]. Then, we can distinguish two topological events:

- **Adding edge**: It consists in adding a new edge to the network at the current date t.
- **Removing edge**: It consists in removing an edge from the network at the current date t.

In order to efficiently construct and maintain tree-based topologies, we use the *DA-GRS*. The latter is a local computation-based model which guarantees that the network remains covered by a spanning forest at any time, in which (1) no cycle can possibly occur, (2) every node belongs to a tree (an isolated node belongs to a tree with a single node which is the root), and (3) there is always exactly one root in every tree. The *DA-GRS* is based on three rules which are presented in Fig. 3:

- **R1: Merging rule**. Whenever two roots (nodes labeled R) arrive at the endpoints of the same edge, one of them destroys its token and selects the other as parent. As a result of this rule, the two trees merge.
- **R2: Circulation rule**. If there is no possible merging, a node in the state R (has the token) passes the token to one of its neighbors in the tree (child) which becomes the new root.

Fig. 3 Three rules for the
DA-GRS

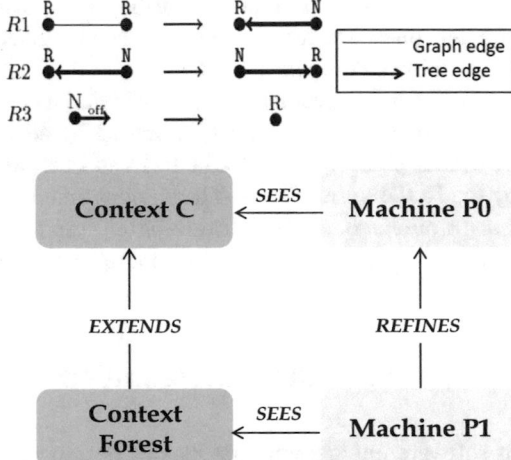

Fig. 4 The refinement
strategy of the proposed
pattern

- **R3**: **Regeneration rule**. Whenever an edge of the tree disappears, the node on the
  child side (labeled *N*) does not possess the token. In this case, there will exist a
  tree without a token. Then, the node must regenerate a token (i.e., it becomes a
  root).

An example of execution sequence of these rules is available in [8].

  In this work, we suppose that the incrementation of time from a date *t* to a date
t+1 is done after: (1) at least one appearance and/or disappearance of an edge is performed in the network and (2) each connected component of the network is covered
by a spanning tree.

4   Formal Development of the Pattern

As mentioned earlier, the specification of our pattern is performed with the Event-B
method and done with the RODIN platform. An Event-B development is based on the
correct-by-construction approach which supports an incremental process controlled
by the refinement of models. We note that two basic levels are necessary to build a
correct pattern as shown in Fig. 4.

  The first model P0 (an abstract machine): We can notice only the appearance
of new edges and disappearance of other edges from a graph G_i at a date t_i to the
following graph G_{i+1} at a date t_{i+1}. The system time is initialized to zero ($t = 0$). At
this date, no topological event (events which are modifying the topology) has been
performed. The incrementation of time is done if one or several topological events
(adding edge and/or removing edge) have been produced. Formally, we define three
events:

- *Adding_Edge*: This event is activated when an edge does not belong to the graph at a current date t. As a result, this edge will be added to the graph.
- *Removing_Edge*: It consists in removing an existing edge from the graph at a current date t.
- *Incrementing_Time*: This event is activated when one or several topological events (*Adding_Edge* and/or *Removing_Edge*) occur in the network.

The second model P1: Once the machine of the first level has been specified and proven, it can be refined in order to build and maintain a forest of spanning trees in dynamic networks. In fact, we introduce labels of nodes to specify the *DA-GRS* rules. Formally, we add three events (*Merging_Rule*, *Regeneration_Rule* and *Circulation_Rule*) and we refine the events specified in *P0* to take into consideration the local label modification. At this level, we indicate that the incrementation of time can take place if each connected component of the network is covered by a spanning tree. In fact, we suppose that the algorithm of building and maintaining a spanning forest (*DA-GRS* rules) acts as an "*observer*" that knows when a spanning tree is formed in each component. This kind of detection is called *observed termination detection* [14].

With these machines, contexts are required with a particular definition in the specification. The first one is the context C which defines basic properties of the network. The second one is the context *Forest*. It is defined as an extension of the context C. It specifies elements of a tree and includes node labels that describe the *DA-GRS* rules.

4.1 Formal specification of the different contexts of the pattern

4.1.1 The Context C

A graph is modeled by a set of nodes called V. In our work, we have supposed that a dynamic graph is composed of stable nodes and variable edges. For this reason, we define V in the context as an abstract set. By means of the *axm1* (*axm1: finite* (V)), we indicate that the number of nodes in the network is finite. Moreover, we introduce a constant, called tn, which represents the final system date. This constant is an integer different to the start date of the system (*axm2*: $tn \in \mathbb{N}1$).[1]

4.1.2 The Context *Forest*

A tree can be defined as an acyclic and connected subgraph. In order to specify a tree, we have to define a node r ($r \in V1; V1 \subseteq V$) which is the root of the tree and a parent function t. Otherwise, each node has a unique parent node, except for the

---

[1]$\mathbb{N}1'$ denotes the set of positive natural numbers: $\mathbb{N}1 = \mathbb{N}\backslash\{0\}$.

root. For more information about tree building, the reader can see [5]. Formally, we obtain the following Event-B definition: $t \in V1 \setminus \{r\} \rightarrow V1$. A tree is an acyclic subgraph. A cycle c in a finite graph t built on a set $V1$ is a subset of $V1$ whose elements are members of the inverse image of c under t, formally, $c \subseteq t^{-1}[c]$. In order to guarantee the non-existence of a cycle in a tree, we must prove that the set c is equal to the empty set. Formally, we describe this property in the following way: $\forall c \cdot (c \subseteq V1 \wedge c \subseteq t^{-1}[c]) \implies c = \emptyset$. Finally, we introduce the constant *trees* to be the set of all trees (with root r) of the graph g. Also, we add some requirement properties: *trees* is non-empty set of possible trees on the graph (axm2) and it is finite (axm3). We specify label nodes as a set called *LN_tree* (axm4). In fact, each node is *labeled R or N*.

```
context Forest
extends C
sets    LN_tree
constants    trees, R, N
axioms
    axm1 : trees = {t, r, V1 · V1 ⊆ V ∧ r ∈ V1 ∧ t ∈ (V1 \ {r}) → V1∧
    (∀c · c ⊆ V1 ∧ c ⊆ (t⁻¹[c]) ⟹ c = ∅)|t}
    axm2 : trees ≠ ∅
    axm3 : finite(trees)
    axm4 : partition(LN_tree, {R}, {N})
end
```

4.2 Formal specification of the different machines of the pattern

4.2.1 The Initial Model (Machine P0)

At this level, a network can be formally modeled as a simple and undirected graph g where nodes denote processors and edges denote direct communication links (inv1). An undirected graph means that there is no distinction between two nodes associated with each edge (inv2). A graph is simple if it has zero or one edge between any two nodes and no edge starts and ends at the same node (inv3). The domain restriction "$V \lhd id$" is a subset of the relation id that contains all of the pairs whose first element is in V. The identity relation id maps every element to itself. Moreover, we introduce a new variable called "*change*" (inv5). If one topological event has been produced, "*change*" is equal to "1", otherwise "*change*" is equal to "0". By means of the invariant *inv6*, we indicate that if the current date t is strictly greater than "0" and "*change*" is equal to "0", then the graph does not undergo any topological event ($g(t) = g(t-1)$). However, if the graph does not remain stable at the date t ($g(t) \neq g(t-1)$), then the variable "*change*" is equal to "1" (inv7). Formally, the invariant specification of *P0* is done as follows:

$$
\begin{aligned}
&inv1 : g \in 0..t \rightarrow \mathbb{P}(\mathbb{V} \times \mathbb{V}) \\
&inv2 : \forall ti \cdot ti \in dom(g) \Longrightarrow g(ti) = (g(ti))^{-1} \\
&inv3 : \forall ti \cdot ti \in dom(g) \Longrightarrow (\mathbb{V} \lhd id) \cap g(ti) = \emptyset \\
&inv4 : t \in \mathbb{N} \wedge t \leq tn \\
&inv5 : change \in \{0, 1\} \\
&inv6 : t > 0 \wedge change = 0 \Longrightarrow g(t) = g(t-1) \\
&inv7 : t > 0 \wedge g(t) \neq g(t-1) \Longrightarrow change = 1
\end{aligned}
$$

Initially, the date is equal to zero ($t = 0$). Also, the variable *change* is equal to zero which means that no topological event has been produced. At this abstract level, we define three events:

- **Event Adding_Edge**: This event is activated when an edge "$x \mapsto y$" between the nodes x and y does not belong to the graph g at the current date t (grd1, grd2, and grd3) and "t" is different to the final date "tn". As a result, this edge will be added to $g(t)$. To respect the invariant *inv2*, we add both "$x \mapsto y$" and "$y \mapsto x$" to $g(t)$ (act1). Moreover, the variable *change* takes the value "1" (act2) to indicate that a topological event has been produced.
- **Event Removing_Edge**: An edge has been removed at the current date t if it is present at the date t (grd1) and "t" is different to the final date "tn". Then, it disappears at the date t (act1) and the variable *change* receives the value "1" (act2).
- **Event Incrementing_Time**: Once the variable *change* takes the value "1" (grd2), the event *Incrementing_Time* can be triggered. In the guard component, we verify that the current date t is strictly lower than the final system date tn. In the action component, we increment the time to $t + 1$ and we set the graph at the date $t + 1(g(t + 1))$ to the graph $g(t)$ (act1). To do this, we use the symbol ": |" which represents a non-deterministic assignment in Event-B. In addition, we reset the variable *change* (act2). Then, we have no topological change at the time $t + 1$.

```
Adding_Edge
any   x, y
where
   grd1 : x ↦ y ∈ V × V
   grd2 : x ↦ y ∉ g(t) ∧ y ↦ x ∉ g(t)
   grd3 : x ≠ y
   grd4 : t ≠ tn
then
   act1 : g(t) := g(t) ∪ {x ↦ y, y ↦ x}
   act2 : change := 1
end
```

```
Incrementing_Time
where
   grd1 : t < tn
   grd2 : change = 1
then
   act1 : g, t : |t′ = t + 1 ∧
g′ = g ∪ {t′ ↦ g(t′)}
   act2 : change := 0
end
```

```
Removing_Edge
any
   x, y
where
   grd1 : x ↦ y ∈ g(t)
   grd2 : t ≠ tn
then
   act1 : g(t) := g(t) \ {x ↦ y, y ↦ x}
   act2 : change := 1
end
```

4.2.2   The Second Model (Machine P1)

We specify the machine *P1* by adding two variables:

Trees_t: It is defined as a total function which assigns a set of disjoint trees $\mathbb{P}(trees)$ to each date from $(0..t)$ (see inv1).

lab: It is defined as a total function which assigns a label R or N from *LN_tree* to each node at a date from $(0..t)$ (see inv2).

$$inv1 : Trees\_t \in 0..t \rightarrow \mathbb{P}(trees)$$
$$inv2 : lab \in (V \times (0..t)) \rightarrow LN\_tree$$

The addition of these two variables involves adding new properties. We have formalized these properties in form of Event-B invariants as following:

- **There is no intersection between the nodes of disjoint trees**: In the invariant *inv3*, we ensure that for all disjoint trees *tr1* and *tr2* at a date *ti* $(ti \in (0..t))$, the intersection between the nodes of *tr1* and *tr2* is empty.
  inv3 : $\forall ti, tr1, tr2 \cdot ti \in dom(Trees\_t) \wedge tr1 \in Trees\_t(ti) \wedge tr2 \in Trees\_t(ti) \wedge$ $tr1 \neq tr2 \Longrightarrow (dom(tr1) \cup ran(tr1)) \cap (dom(tr2) \cup ran(tr2)) = \emptyset$

- **Each disjoint tree has only one root labeled R and all the other nodes are labeled N**: This constraint is ensured by the invariant *inv4* where for all disjoint tree *tr* at a date *ti* $(ti \in (0..t))$, one node "x" of the tree *tr* is labeled R and all the other nodes "y" are labeled N.
  inv4 : $\forall ti, tr \cdot ti \in dom(Trees\_t) \wedge tr \in Trees\_t(ti) \Longrightarrow (\exists x \cdot (x \mapsto ti) \in dom(lab)$ $\wedge lab(x \mapsto ti) = R \wedge (\forall y \cdot y \in (dom(tr) \cup ran(tr)) \setminus \{x\} \wedge (y \mapsto ti) \in dom(lab) \Longrightarrow$ $lab(y \mapsto ti) = N))$

- **A node which does not belong to any disjoint tree (it can belong to graph edges) is labeled R**: This property is expressed by the invariant *inv5* where we ensure that, at a date *ti* $(ti \in (0..t))$, a node which does not belong to any disjoint tree is labeled R. In fact, this node forms a tree of its own and it is the root of this tree.
  inv5 : $\forall ti, x \cdot ti \in dom(Trees\_t) \wedge x \in V \wedge (\{\{x\}\} \cap \{tr.tr \in Trees\_t(ti) | dom(tr)$ $\cup ran(tr)\} = \emptyset) \Longrightarrow lab(x \mapsto ti) = R$

Initially, all the nodes are labeled R at the date "$t = 0$". Otherwise, every node is considered as a tree with a single node and it is the root of this tree. Then, the set of disjoint trees is empty *(Trees_t(0)= $\emptyset$)*. In order to specify the *DA-GRS* rules, we introduce three events: *Merging_Rule, Regeneration_Rule*, and *Circulation_Rule*. Due to space limitation, we only detail the specification of the event *Circulation_Rule*. This event specifies the rule *R3* displayed in Fig. 3. The specification of this event is done as follows:

In the guard component, we define *grd1* to verify the presence of an edge "$y \mapsto x$" in the graph *g* at the current date *t*. More precisely, the edge "$y \mapsto x$" belongs to a disjoint tree *tr* at the date *t* (grd2 and grd3). Moreover, we indicate in the *grd4* that the nodes *y* and *x* are, respectively, labeled N and R at the date *t*. In the action component, we define two actions: By means of the action *act1*, we update the set of disjoint trees at the date *t*. In fact, the disjoint tree *tr* is replaced by the tree

$(tr \setminus \{y \mapsto x\}) \cup \{x \mapsto y\}$. To do so, we use the overriding operator "$\Leftarrow$" in Event-B. Also, "$lab : |lab' = lab \Leftarrow \{(x \mapsto t) \mapsto N, (y \mapsto t) \mapsto R\}$" ($act2$) means that the new labels of the node x and y at the date t are, respectively, N and R.

```
Circulation_Rule
any   x, y, tr
where
    grd1 : y ↦ x ∈ g(t)
    grd2 : t ∈ dom(Trees_t) ∧ tr ∈ Trees_t(t)
    grd3 : y ↦ x ∈ tr
    grd4 : (x ↦ t) ∈ dom(lab) ∧ (y ↦ t) ∈ dom(lab)
         ∧lab(x ↦ t) = R ∧ lab(y ↦ t) = N
then
    act1 : Trees_t : |Trees_t′ = Trees_t⇐
        {t ↦ (Trees_t(t) \ {tr}) ∪ {(tr \ {y ↦ x}) ∪ {x ↦ y}}}
    act2 : lab : |lab′ = lab ⇐ {(x ↦ t) ↦ N, (y ↦ t) ↦ R}
end
```

Specification of Topological Events

- **Event Adding_Edge**: At this second level, the event *Adding_Edge* presented in the machine *P0* remains unchanged. In fact, the appearance of a new edge in the network at the current date t requires only the addition of the edge, without modifying the labels of nodes.

- **Event Removing_Edge**: We refine the event *Removing_Edge* detailed at the first level by reinforcing the guard component. In fact, we add a new guard *grd2* to indicate that the removed edge does not belong to any disjoint tree at the current date t.

  grd2: $(t \in dom(Trees\_t) \land \forall tr \cdot tr \in Trees\_t(t) \Rightarrow (x \mapsto y \notin tr \land y \mapsto x \notin tr))$

- **Event Incrementing_Time:** We refine the event *Incrementing_Time* presented in the machine *P0* by strengthening the guard component. In fact, we add a new guard *grd3* to indicate that each connected component at the date t is covered by a spanning tree. So, two neighboring nodes x and y of the graph should belong to the same spanning tree.

  grd3: $\forall tr, x, y \cdot tr \in Trees\_t(t) \land x \in (dom(tr) \cup ran(tr)) \land (x \mapsto y \in g(t) \lor y \mapsto x \in g(t)) \Longrightarrow y \in (dom(tr) \cup ran(tr))$

  In addition, we reinforce the action *act1* to indicate that the set of disjoint trees at the date $t+1$ is equal to *Trees_t(t)*. Also, the labels of nodes at the date $t+1$ are equal to the current labels at the date t.

  act1: $g, t, Trees\_t, lab : |t' = t + 1 \land g' \in 0 .. t' \rightarrow \mathbb{P}(V \times V) \land g' = g \Leftarrow \{t' \mapsto g(t)\} \land Trees\_t' \in 0 .. t' \rightarrow \mathbb{P}(tree) \land Trees\_t' = Trees\_t \Leftarrow \{t' \mapsto Trees\_t(t)\} \land lab' \in (V \times (0..t')) \rightarrow LN\_tree \land lab' = lab \Leftarrow \{y \cdot y \in V | (y \mapsto t') \mapsto lab(y \mapsto t)\}$

5 Case Study: Leader Election Algorithm

To illustrate the proposed pattern, we present an example of a leader election algorithm which operates on tree-based topologies, encoded by the local computations model. The main objective of this section is to demonstrate how our pattern can be used and incorporated during development to specify the election algorithm.

Fig. 5 Relabeling rules of
the leader election algorithm

5.1 Algorithm Presentation

Consider a tree where each node is initially labeled by its degree (d). The node degree is the number of tree edges connected to this node. The leader election algorithm is given by two rules r_a and r_b presented in Fig. 5. The first one is a pruning rule which can apply a computation to two adjacent nodes in the network. It consists in cutting a pendant node by giving it the label *Non-Elected* (NE). However, the neighbor node decrements its degree to "d-1" ($d \geq 1$). The label of a node becomes *elected* (E) by the rule r_b if it has no neighbor (d = 0). A run of the algorithm consists in applying the relabeling rules specified by the algorithm until no rule is applicable. In the final configuration, only one node is marked as *elected* "E" and all the other nodes are *non-elected* "NE."

5.2 Using the Proposed Pattern in the Development of the Leader Election Algorithm

In this section, we present the idea of the pattern incorporation into an Event-B development. In fact, we explain how our pattern is used to correctly specify the leader election algorithm. The process is shown in Fig. 6. Generally, the development of a distributed algorithm in Event-B starts with a very abstract model. Then, by successive refinements, we obtain a concrete one that expresses the local behavior of processors in the network. Each refinement level is defined by an Event-B machine.

We follow the different steps to refine and incorporate the proposed pattern during the system development:

- **Step 1**: We define a machine *M0* which refines the machine *P0* (❶ in Fig. 6). Then, it includes the events of *P0*. We have to add only one event which specifies the result of the algorithm in one shot and does not describe how the solution is computed. In other words, there is no protocol, only the formal definition of its intended result. The analogy of someone closing and opening their eyes. This event can verify that each connected component at the current date *t* has an elected node.
- **Step 2**: In order to specify the machine *M1*, we refine the machines *M0* and *P1* (❷ and ❸ in Fig. 6). The reader should know that the double refinement is not implemented in RODIN. But, really we have two copies of *M1*. A copy that refines *M0* and another that refines *P0*. In the machine *M1*, we add some details to specify globally the computation of the algorithm result. In fact, we refine the event intro-

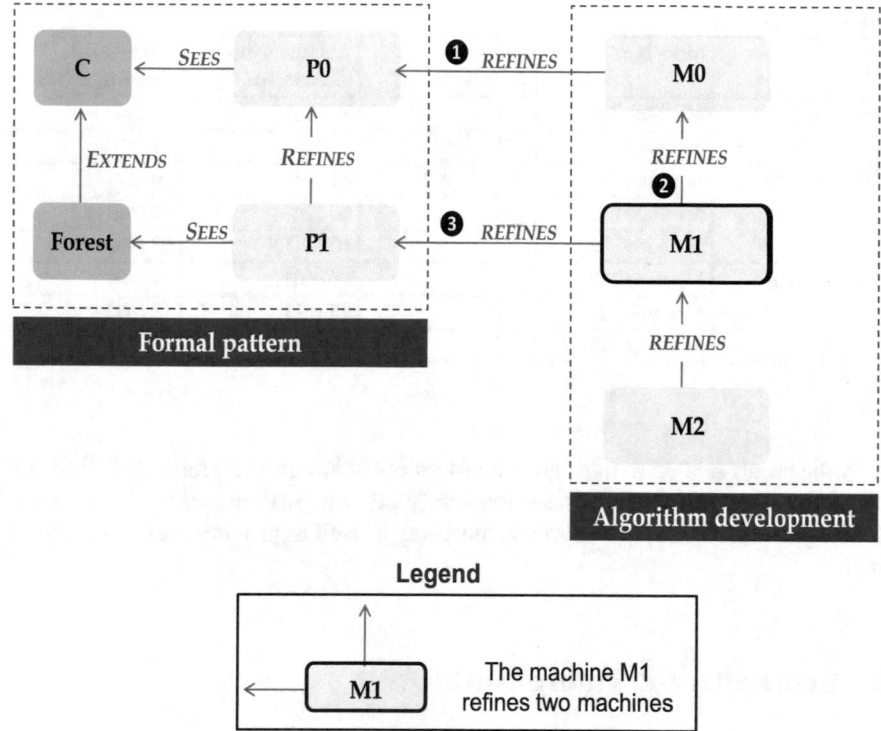

Fig. 6 Using the proposed pattern in the development of the leader election algorithm

duced at the first level to ensure that a node is elected from a spanning tree of each
connected component.

- **Step 3**: We introduce a new machine called *M2* which refines *M1*. *M2* locally
specifies nodes' interactions in order to compute an elected node in each connected
component (spanning tree). It is an application of the relabeling rules r_a and r_b.

5.3 Proof Statistics

One of our objectives in this paper is to reduce the proof efforts. In fact, we aim
to increase automatic proofs and decrease those that need interactive efforts to dis-
charge. Table 1 shows the proof statistics related to the development of the pattern
and the leader election algorithm with and without using the pattern. It includes
the proof obligations generated and discharged by the RODIN platform and those
interactively proved. As we can see, by using our pattern, we notice a significant
reduction of the proofs interactively discharged in *M0* and *M1*. In fact, the machine
M0 is a refinement of *P0*. Also, the machine *M1* refines *M0*. So, we need to prove

Table 1 Proof statistics

	Models	Total proof obligations	Automatically discharged (%)	Interactively discharged (%)
Pattern	Machine P0	41	19 (46)	22 (54)
	Machine P1	105	44 (42)	61(58)
Without pattern	Machine M0	51	42 (82)	9 (18)
	Machine M1	146	50 (34)	96 (66)
	Machine M2	72	10 (14)	62 (86)
With pattern	Machine M0	10	6 (60)	4 (40)
	Machine M1	41	25 (61)	16 (39)
	Machine M2	72	10 (14)	62 (86)

only the proofs related to the algorithm (4 proofs in *M0* and 16 proofs in *M1*). Additionally, we can reuse the pattern to specify other algorithms such as coloration. Consequently, we can save efforts on modeling as well as proving the correctness of models.

6 Conclusion and Future Work

In this paper, we have presented a reuse-based solution for specifying and proving distributed algorithms which operate on tree-based topologies. Our contribution consists in proposing a formal pattern based on the evolving graph model. It relies on the *DA-GRS* to construct and maintain a forest of spanning trees in dynamic networks. To illustrate our pattern, we have presented the leader election algorithm as a case study. The proof statistics show that our solution can save efforts on specifying as well as proving the correctness of distributed algorithms in a forest topology.

As a future work, we plan to illustrate the proposed pattern with other examples of distributed algorithms such as coloration. In addition, we intend to propose a plugin in RODIN which supports the double refinement.

References

1. Abrial, J.R.: Modeling in Event-B: System and Software Engineering. Cambridge University Press (2010)
2. Abrial, J.R., Butler, M., Hallerstede, S., Hoang, T., Mehta, F., Voisin, L.: Rodin: an open toolset for modelling and reasoning in event-b. Int. J. STTT **12**(6), 447–466 (2010)
3. Back, R.J.R.: A calculus of refinements for program derivations. Acta Inform. **25**, 593–624 (1988)

4. Barjon, M., Casteigts, A., Chaumette, S., Johnen, C., Neggaz, Y.M.: Maintaining a spanning forest in highly dynamic networks: the synchronous case. In: The 18th International Conference on Principles of Distributed Systems (PDS), vol. 8878, pp. 277–292. Springer (2014)
5. Cansell, D., Méry, D.: The event-b modelling method: concepts and case studies. In: Logics of Specification Languages, pp. 47–152. Springer, Berlin (2008)
6. Casteigts, A.: Contribution à l'algorithmique distribué dans les réseaux mobiles ad hoc. Ph.D. thesis, Université Sciences et Technologies—Bordeaux I (2007)
7. Casteigts, A., Chaumette, S.: Dynamicity aware graph relabeling systems (DA-GRS), a local computation based model to describe MANET algorithms. In: International Conference on Parallel and Distributed Computing Systems (PDCS), pp. 231–236 (2005)
8. Casteigts, A., Chaumette, S., Guinand, F., Pigné, Y.: Distributed maintenance of anytime available spanning trees in dynamic networks. In: The 12th International Conference on Ad-Hoc Networks and Wireless (ADHOC-NOW), vol. 7960, pp. 99–110. Springer (2013)
9. Casteigts, A., Flocchini, P.: Deterministic algorithms in dynamic networks: problems, analysis, and algorithmic tools. Commissioned by Defense Research and Development Canada, Technical report (2013)
10. Fakhfakh, F., Tounsi, M., Kacem, A.H., Mosbah, M.: A formal pattern for dynamic networks through evolving graphs. In: 12th IEEE/ACS International Conference of Computer Systems and Applications (AICCSA) (2015)
11. Fakhfakh, F., Tounsi, M., Kacem, A.H., Mosbah, M.: A refinement-based approach for proving distributed algorithms on evolving graphs. In: The 25th International Conference on Enabling Technologies: Infrastructure for Collaborative Enterprises (WETICE), pp. 44–49 (2016)
12. Ferreira, A.: On models and algorithms for dynamic communication networks: the case for evolving graphs. In: The 4e Rencontres Francophones sur les Aspects Algorithmiques des Telecommunications (AlgoTel), pp. 155–161. INRIA Press (2002)
13. Floriano, P., Goldman, A., Arantes, L.: Formalization of the necessary and sufficient connectivity conditions to the distributed mutual exclusion problem in dynamic networks. In: IEEE (ed.) The 10th International Symposium on Network Computing and Applications (NCA), pp. 203–210 (2011)
14. Godard, E., Métivier, Y., Tel, G.: Termination Detection of Local Computations. CoRR (2010). arXiv:1001.2785
15. Hoang, T.S., Fürst, A., Abrial, J.R.: Event-b patterns and their tool support. Softw. Syst. Model. 12, 229–244 (2013)
16. Kerchove, F.M.D.: Relabeling algorithms on dynamic graphs. University of Le Havre, Technical report (2012)
17. Leavens, G.T., Abrial, J.R., Batory, D., Butler, M., Coglio, A., Fisler, K., Hehner, E., Jones, C., Miller, D., Peyton-Jones, S., Sitaraman, M., Smith, D.R., Stump, A.: Roadmap for enhanced languages and methods to aid verification. In: The 5th International Conference on Generative Programming and Component Engineering (GPCE), pp. 221–236. ACM (2006)
18. Litovsky, I., Métivier, Y., Sopena, E.: Handbook of graph grammars and computing by graph transformation. In: Chapter Graph Relabelling Systems and Distributed Algorithms, pp. 1–56. World Scientific (1999)
19. Roy, R.: Mobile ad hoc networks. In: Handbook of Mobile Ad-Hoc Networks for Mobility Models, pp. 3–22. Springer, US (2011)
20. Tel, G.: Introduction to Distributed Algorithms. Cambridge University Press (2000)

A Multicriteria Approach for Selecting the Optimal Location of Waste Electrical and Electronic Treatment Plants

Santoso Wibowo and Srimannarayana Grandhi

Abstract This paper presents multicriteria decision-making approach for selecting the optimal location of waste electrical and electronic equipment (WEEE) treatment plants. Hesitant fuzzy set is used to deal with the situations in which the decision maker hesitates among several values to assess the alternatives. A hesitant fuzzy Hamacher geometric operator is proposed for producing a preference value for every waste electrical and electronic equipment treatment plant location alternative across all selection criteria. An example is given for demonstrating the applicability of the proposed multicriteria decision-making approach for selecting the optimal location of WEEE treatment plants.

Keywords Treatment plants · Optimal location · Multicriteria · Selection · Waste electrical and electronic equipment

1 Introduction

The electronics industry is one of the world's largest and fastest growing manufacturing sectors. This increasing consumption of electrical and electronic equipment naturally leads to a high yield of waste electrical and electronic equipment (WEEE). It is found that WEEE is the world's rapidly growing waste stream, which has the growth rate of 3–5% per year [1] and potentially the biggest challenge to sustainability [2]. A research conducted by Achillas et al. [3] found that around 30–50 million tonnes of WEEE is disposed each year globally. On top of that, producing and discarding more electronic products will lead to more mining, more fossil fuel extraction, and more refining, with all of the direct and secondary environmental and health impacts that come with these processes [4].

S. Wibowo (✉) · S. Grandhi
School of Engineering & Technology, CQUniversity, Melbourne, Australia
e-mail: s.wibowo1@cqu.edu.au

S. Grandhi
e-mail: s.grandhi@cqu.edu.au

© Springer International Publishing AG 2018
R. Lee (ed.), *Computer and Information Science*, Studies in Computational
Intelligence 719, DOI 10.1007/978-3-319-60170-0_10

The potential for negative environmental impact resulting from the treatment of these wastes is high due to the presence of hazardous substances within the waste stream [5]. Therefore, government agencies, businesses, and the public are becoming increasingly interested in the alternative management of industrial products when those reach the end of their useful life. In order to efficiently manage WEEE products at the end of their useful life, it is critical for organizations to select the optimal location of WEEE treatment plant.

Selecting the optimal location of WEEE treatment plant alternatives is complex and challenging. This is due to the involvement of multiple decision makers in evaluating the available alternatives with respect to multiple, often conflicting criteria, and the presence of subjectiveness and imprecision of the decision-making process [6].

Much research has been conducted for dealing with the WEEE treatment plant location selection problem [4, 5, 7]. For example, Queiruga et al. [4] apply the PROMETHEE (Preference Ranking Organization METHod for Enrichment Evaluations) approach for the evaluating and selecting the location of WEEE recycling plants in Spain. The required information for the evaluation and selection process is gathered and analyzed through the use of a structured questionnaire that is filled in by the experts. This approach is then applied to assess the most suitable location for WEEE recycling plants.

Kim et al. [5] present a hybrid approach for solving WEEE decision problem with the combination of the analytical hierarchy process (AHP) and the Delphi approach. Appropriate evaluation criteria were derived using the Delphi approach to assess the potential selection and priority among the available alternatives. The weightings from the AHP are calculated to identify the priorities of alternatives.

Yuksel [7] applies the AHP for the selection of WEEE collection center location. With the use of this approach, the evaluation and selection problem is formulated in a hierarchical structure, and pairwise comparison is used for determining the performance of each WEEE location with respect to each criterion and the importance of the evaluation and selection criteria. The overall performance of each WEEE location across all criteria is determined based on the utility theory.

These approaches, however, are found to have various shortcomings including (a) the failure to adequately handle the subjectiveness and imprecision of the decision-making process, (b) tedious mathematical computation required, and (c) cognitively very demanding on the decision maker.

To overcome the limitations of these existing approaches, this paper presents a multicriteria decision-making approach for selecting the optimal location of WEEE treatment plants. Hesitant fuzzy set is used to deal with the situations in which the decision maker hesitates among several values to assess the alternatives. A hesitant fuzzy Hamacher geometric operator is proposed for producing a preference value for every WEEE treatment plant location alternative across all selection criteria. An example is given for demonstrating the applicability of the proposed multicriteria decision-making approach for selecting the optimal location of WEEE treatment plants.

2 The WEEE Treatment Plant Location Selection Problem

The evaluation and selection of WEEE treatment plant location alternatives with respect to a set of specific criteria is complex [4]. This is due to the presence of the multidimensional nature of the evaluation process and the presence of vagueness of the decision-making process [6, 8]. To effectively deal with this problem, an overall evaluation of individual WEEE treatment plant location alternatives is desirable.

In order to measure the performance of the available WEEE treatment plant location alternatives, it is important to firstly define the suitable criteria for ensuring that the evaluation and selection process produces an accurate and effective result. This is because not every criterion is relevant to the specific requirements [3–5, 8–13].

Much research has been done on identifying the relevant criteria for selecting the optimal location of WEEE treatment plant alternatives [4, 8]. Queiruga et al. [4] state that land costs, facility access, availability of labor, and proximity to inhabitant are important criteria for evaluating the most suitable WEEE treatment plant location. Meanwhile, Yuksel [7] and Hokkanen and Salminen [8] believe that operating cost, effect on the environment, and health concerns are criteria for evaluating the most suitable WEEE treatment plant location. Chang and Wang [9] state that minimum net cost, minimum traffic congestion, and minimum air pollution are the critical criteria for evaluating the most suitable WEEE treatment plant location. Chambal et al. [10] believe that the most suitable WEEE treatment plant location should have a minimum operating cost, a minimum negative impact on the environment, and comply with the state, federal, and other regulations. Cheng et al. [11] point out that operating costs, effects on environment, and effects on people should be taken into account while evaluating and selecting the most suitable WEEE treatment plant location. At the same time, Adamides et al. [12] state that low transportation cost and minimum social resistances should also be considered while evaluating the most suitable WEEE treatment plant location. Liu et al. [13] state that the most suitable WEEE treatment plant location should be in close proximity to building materials for landfill construction and operation and far away from the dense population.

A review of the related literature leads to the classification of the critical criteria into (a) capital cost, (b) accessibility, (c) environmental impact, (d) site capacity, and (e) infrastructure. Figure 1 shows the hierarchical structure of the WEEE treatment plant location selection problem.

Capital cost (C_1) refers to both operating cost and fixed cost. Operating cost covers the movement of WEEE and labor charges. Fixed cost refers to capital investments made on acquiring land and other resources for the establishment of recycling plant [14]. Accessibility (C_2) refers to the level of accessibility to the facility. The close proximity of the facility from the customers is considered vital for transportation. This is an important criterion as it impacts on the collection of WEEE [15]. Environmental impact (C_3) reflects on potential effects of the WEEE to the environment. WEEE consists of both hazardous and non-hazardous materials.

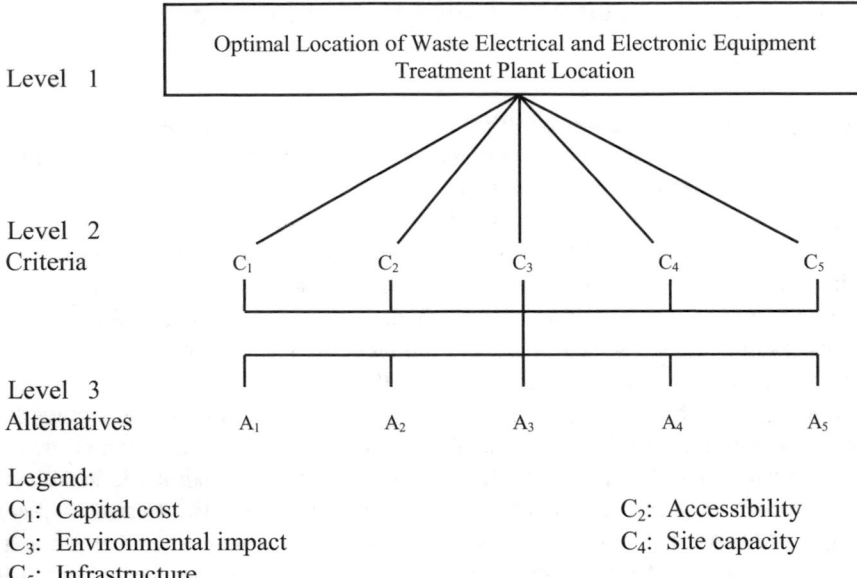

Legend:
C₁: Capital cost   C₂: Accessibility
C₃: Environmental impact   C₄: Site capacity
C₅: Infrastructure

$A_i (i = 1, 2, \ldots, n)$: Waste electrical and electronic equipment treatment plant location alternatives.

Fig. 1 Hierarchical structure of waste electrical and electronic equipment treatment plant location selection problem

While recycling these materials, it is important to consider the impact of recycling methods on surrounding environment and mitigate environmental pollution [5, 14]. Site capacity (C_4) refers to the capacity and size of the WEEE treatment plant location to recycle a large amount of WEEE. These WEEEs may include small and large household appliances, IT equipment, consumer equipment, lighting equipment, industrial tools, medical devices, automatic dispensers, monitoring and control equipment [14]. Infrastructure (C_5) reflects on the access to roads, buildings, transportation and communication network, electricity, and water required for the smooth operation of the WEEE treatment plant location. Kiddee et al. [16] point out that only around 10% of WEEE is recycled due to lack of a safe recycling infrastructure. A proper infrastructure is the backbone for WEEE collection, recycling, and transportation of recovered materials.

3 Multicriteria Decision-making Approach

Evaluating and selecting the optimal location of WEEE treatment plant alternatives involves in (a) discovering all the WEEE treatment plant location alternatives, (b) identifying the selection criteria, (c) assessing the WEEE treatment plant location alternatives' performance ratings and the criteria, and (d) selecting the best WEEE treatment plant location alternative [17].

Formulated as a multicriteria decision-making problem, the evaluation and selection for available WEEE treatment plant location alternatives generally starts with (a) identifying different WEEE alternatives, (b) selecting relevant criteria for evaluation, (c) evaluating each available WEEE alternative and their performance ratings along with the criteria weights, (d) preparing overall preference values for every WEEE alternative by multiplying the alternative ratings and criteria weights, and finally (e) choosing the most appropriate WEEE alternative in a given situation [6, 18].

In many practical problems, it is often difficult to define the membership grade of an element because of a set of possible membership values. This situation is very common in the multicriteria group decision-making problem when decision makers are not agreeing on the same membership grade for an element. To deal with this issue, hesitant fuzzy sets [19] were introduced, whereby these fuzzy sets are described by a membership function which is represented by a set of possible values between zero and one.

For a multicriteria decision-making problem with hesitant fuzzy information, let A_i ($i = 1, 2, ..., n$), C_j ($j = 1, 2, ..., m$) be the set of criteria with the criteria weight vectors $w = (w_1, w_2, ..., w_m)^T$ where $w_j > 0$ ($j = 1, 2, ..., m$) and $\sum_{j-1}^{m} w_j = 1$. Here, the performance rating value of alternative A_i ($i = 1, 2, ..., n$) with respect to criteria C_j ($j = 1, 2, ..., m$) is expressed by a fuzzy hesitant element $h_{ij} = \bigcup_{\gamma_{ij} \in h_{ij}} \{\gamma_{ij}\} \in H$, which is a set of possible evaluated values of alternative $a_i \in A$ with respect to criteria $c_j \in C$. Thus, a hesitant fuzzy decision-making matrix H can be represented as

$$H = \begin{bmatrix} h_{11} & h_{12} & \cdots & h_{1m} \\ h_{21} & h_{22} & \cdots & h_{2m} \\ \cdots & \cdots & \cdots & \cdots \\ h_{n1} & h_{n2} & \cdots & h_{nm} \end{bmatrix} \quad (1)$$

Normally, there are two types of criteria to be considered in multicriteria decision-making problems namely cost criteria and benefit criteria. If all the criteria C_j ($j = 1, 2, ..., m$) are of the same type, then the criteria values do not need to be normalized. If there are benefit criteria and cost criteria in the multicriteria decision-making problem, the criteria values of the cost type can be transformed into criteria values of the *benefit type by using (2)*.

$$r_{ij} = \begin{cases} h_{ij}, & \text{for benefit attribute } c_j \\ (h_{ij})^c, & \text{for cost attribute } c_j \end{cases}, i = 1, 2, ..., n; j = 1, 2, ..., m \quad (2)$$

where $(h_{ij})^c$ is the complement of h_{ij} such that $(h_{ij})^c = \bigcup_{\gamma_{ij} \in h_{ij}} \{1 - \gamma_{ij}\}$. In this case, the hesitant fuzzy decision matrix $H = (h_{ij})_{m \times n}$ can be transformed into a corresponding hesitant fuzzy decision matrix $R = (r_{ij})_{m \times n}$.

To obtain the most suitable alternative, the hesitant fuzzy Hamacher weighted geometric (HFHWG) [20] operator is introduced for dealing with the hesitant fuzzy multicriteria decision-making problem. The procedure for HFHWG operator involves the following steps:

Step 1. The decision maker provides his/her performance rating values of alternative $A_i \in A$ with respect to criteria $c_j \in C$, which are expressed by hesitant fuzzy elements h_{ij} ($i = 1, 2, ..., n, j = 1, 2, ..., m$).

Step 2. The hesitant fuzzy decision-making matrix $D = (h_{ij})_{m \times n}$ is transformed into a normalized matrix $R = (r_{ij})_{m \times n}$ based on (2).

Step 3. Aggregate all the hesitant fuzzy values r_{ij} ($j = 1, 2, ..., m$) into a global value r_i by using the HFHWG operator as in (3).

$$r_i = HFHWG_\zeta(r_{i1}, r_{i2}, ..., r_{in}) = \overset{m}{\underset{j=1}{\otimes}} (r_{ij})^{w_j}$$

$$= \bigcup_{\gamma_{i1} \in r_{i1}, \gamma_{i2} \in r_{i2}, ..., \gamma_{in} \in r_{in}} \left\{ \frac{\zeta \prod_{j=1}^{m} (\gamma_{ij})^{w_j}}{\prod_{j=1}^{m} (1 + (\zeta-1)(1-\gamma_{i1})^{w_j} + (\zeta-1) \prod_{j=1}^{m} (\gamma_{i1})^{w_j}} \right\}$$

$$(3)$$

Step 4. Compute the scores $s(r_i)$ of the global hesitant fuzzy r_i ($i = 1, 2, ..., n$).

$$s(r_i) = \sum_{\gamma \in h} \gamma / \delta(h) \qquad (4)$$

where $\delta(h)$ is the number of elements in h.

Step 5. Rank all the alternatives Ai and select the most suitable alternative based on the score value of $s(r_i)$.

The larger the $s(r_i)$ value, the more preferred the alternative A_i.

4   An Example

This section presents the applicability of the proposed multicriteria decision-making approach described in Sect. 3 for selecting the optimal WEEE treatment plant location alternative.

The production output of electrical and electronic manufacturing industry in Cambodia has risen dramatically in the last few years due to the technological advances and customers' demand. As a result of this, WEEE has become a problem of great concern as it is estimated that WEEE generation potential ranges from 6792 metric tons in 2008 to 22 443 metric tons in 2019 [16]. This increasing concern about WEEE has led the government agency to study an optimal location of a WEEE treatment plant alternative for the management of WEEE.

A project team consisting of three decision makers is formed to assess that current WEEE collection, transportation, treatment, and disposal. A detailed study is conducted through the distribution of a questionnaire and the conduct of face-to-face interviews to managers from different companies and the related authorities in the manufacturing industry. Both the questionnaire and the interviews have managed to successfully gather information regarding the important criteria for evaluating and selecting the optimal location of the WEEE treatment plant.

Four WEEE treatment plant location alternatives have to be evaluated with the aim of finding the best performing WEEE treatment plant location.

Based on the information provided on the five relevant criteria discussed in the previous section and the four WEEE treatment plant location alternatives, we utilize the proposed approach presented in the Sect. 3 for selecting the optimal WEEE treatment plant location alternatives.

Step 1. During the evaluation process, it is necessary for the decision makers to provide their own different preferences to which the alternative A_i satisfies C_j. Based on their agreement, the weight vector of the criteria is given as $W = (0.1, 0.15, 0.2, 0.15, 0.4)^{\mathrm{T}}$. Then, the decision matrix $D = (h_{ij})_{m \times n}$ can be expressed by the hesitant fuzzy sets shown as in Table 1.

Step 2. The hesitant fuzzy decision-making matrix $D = (h_{ij})_{m \times n}$ is transformed into a normalized matrix $R = (r_{ij})_{m \times n}$, based on (2).

Step 3. Let $\zeta = 0.5$, and aggregate all of the preference values h_{ij} ($j = 1, 2, 3, 4, 5$) in the ith line of the decision matrix H, by using (2), into global values r_i of alternative A_i ($i = 1, 2, 3, 4, 5$).

Step 4. Calculate the score values $s(r_i)$ ($i = 1, 2, 3, 4, 5$) of r_i. Table 2 shows the results.

Step 5. Rank all WEEE treatment plant location alternatives A_i ($i = 1, 2, 3, 4, 5$) in accordance with the scores $s(r_i)$ of the overall hesitant fuzzy preference values.

By using the multicriteria decision-making approach illustrated in the previous section, the overall preference value for each WEEE treatment plant location alternative across all the criteria can be calculated. The overall preference values of

Table 1 Hesitant fuzzy decision matrices

	C_1	C_2	C_3	C_4	C_5
A_1	{0.7, 0.9}	{0.4, 0.6}	{0.2, 0.4}	{0.2, 0.4, 0.8, 0.9}	{0.2, 0.3, 0.6}
A_2	{0.4, 0.7}	{0.5, 0.7}	{0.3, 0.6, 0.7}	{0.3, 0.6}	{0.1, 0.2, 0.4}
A_3	{0.2, 0.4, 0.6}	{0.2, 0.5, 0.7, 0.8}	{0.2, 0.5, 0.6, 0.8}	{0.2, 0.4, 0.8}	{0.6, 0.9}
A_4	{0.3, 0.6, 0.7, 0.9}	{0.1, 0.4, 0.8}	{0.3, 0.5, 0.6, 0.9}	{0.5, 0.6, 0.9}	{0.5, 0.9}
A_5	{0.8}	{0.2, 0.5}	{0.3, 0.8, 0.9}	{0.1, 0.3}	{0.2, 0.6}

Table 2 Score values and the rankings of the alternatives

Alternatives	Values	Rankings
A_1	0.5348	2
A_2	0.5692	1
A_3	0.4759	3
A_4	0.3825	5
A_5	0.4281	4

the WEEE treatment plant location alternatives and their corresponding rankings are shown in Table 2. Alternative A_2 is the most suitable WEEE treatment plant location for selection as it has the highest score value of 0.5692.

It can be observed that the multicriteria decision-making approach is a useful and practical tool for decision makers who need to evaluate and select available WEEE treatment plant locations and make critical decisions regarding the optimal location for the WEEE treatment plant.

5 Conclusion

This paper has presented multicriteria decision-making approach for selecting the optimal location of WEEE treatment plant location. Hesitant fuzzy set is used to deal with the situations in which the decision maker hesitates among several values to assess the alternatives. A hesitant fuzzy Hamacher geometric operator is proposed for producing a preference value for every WEEE treatment plant location alternative across all selection criteria. An example is presented that clearly demonstrates the potentiality, the applicability, and the simplicity of the multicriteria decision-making approach in providing a multicriteria decision aid to the decision maker during the selection process.

References

1. Afroz, R., Masud, M.M., Akhtar, R., Bt Duasa, J.: Survey and analysis of public knowledge, awareness and willingness to pay in Kuala Lumpur, Malaysia—A case study on household WEEE management. J. Clean. Prod. **52**, 185–193 (2013)
2. Qu, Y., Zhu, Q., Sarkis, J., Yong Geng, Y., Zhong, Y.: A review of developing an e-wastes collection system in Dalian, China. J. Clean. Prod. **52**, 176–184 (2013)
3. Achillas, Ch., Vlachokostas, Ch., Moussiopoulos, N., Banias, G.: Decision support system for the optimal location of electrical and electronic waste treatment plants: a case study in Greece. Waste Manag. **30**, 870–879 (2010)
4. Queiruga, D., Walther, G., Gonzalez-Benito, J., Spengler, T.: Evaluation of sites for the location of WEEE recycling plants in Spain. Waste Manag. **28**, 181–190 (2008)
5. Kim, M., Jang, Y.C., Lee, S.: Application of Delphi-AHP methods to select the priorities of WEEE for recycling in a waste management decision-making tool. J. Environ. Manag. **128**, 941–948 (2013)

6. Wibowo, S., Deng, H.: Multi-criteria group decision making for evaluating the performance of e-waste recycling programs under uncertainty. Waste Manag. **40**, 127–135 (2015)

7. Yuksel, H.: An analytical hierarchy process decision model for e-waste collection center location selection. In: International Conference on Computers and Industrial Engineering, France, 6–9 July 2009

8. Hokkanen, J., Salminen, P.: Locating a waste treatment facility by multicriteria analysis. J. Multi-Criteria Decis. Anal. **6**, 175–184 (1997)

9. Chang, N.B., Wang, S.F.: The development of an environmental decision support system for municipal solid waste management. Comput. Environ. Urb. Syst. J. **20**, 201–212 (1996)

10. Chambal, S., Shoviak, M., Thal Jr., A.E.: Decision analysis methodology to evaluate integrated solid waste management alternatives. Environ. Model. Assess. **8**, 25–34 (2003)

11. Cheng, S., Chan, C.W., Huang, G.H.: Using multiple criteria decision analysis for supporting decisions of solid waste management. J. Environ. Sci. Health A Toxic/Hazard. Subst. Environ. Eng. **37**, 975–990 (2002)

12. Adamides, E.D., Mitropoulos, P., Giannikos, I., Mitropoulosm, I.: A multi-methodological approach to the development of a regional solid waste management system. J. Oper. Res. Soc. **60**, 758–770 (2009)

13. Liu, H.C., You, J.X., Chen, Y.Z., Fan, X.J.: Site selection in municipal solid waste management with extended VIKOR method under fuzzy environment. Environ. Earth Sci. **72**, 4179–4189 (2014)

14. Osibanjo, O., Nnorom, I.C.: The challenge of electronic waste (e-waste) management in developing countries. Waste Manag. Res. **25**, 489–501 (2007)

15. Townsend, T.G.: Environmental issues and management strategies for waste electronic and electrical equipment. J. Air Waste Manag. Assoc. **61**, 587–610 (2011)

16. Kiddee, P., Naidu, R., Wong, M.H.: Electronic waste management approaches: an overview. Waste Manag. **33**, 1237–1250 (2013)

17. Wibowo, S., Deng, H.: Consensus-based decision support for multicriteria group decision making. Comput. Ind. Eng. **66**, 625–633 (2013)

18. Wibowo, S., Deng, H.: Intelligent decision support for effectively evaluating and selecting ships under uncertainty in marine transportation. Expert Syst. **39**, 6911–6920 (2012)

19. Torra, V.: Hesitant fuzzy sets. Int. J. Intell. Syst. **25**, 529–539 (2010)

20. Tan, C., Yi, W., Chen, X.: Hesitant fuzzy Hamacher aggregation operators for multicriteria decision making. Appl. Soft Comput. **26**, 325–349 (2015)

Localization Strategy for Island Model Genetic Algorithm to Preserve Population Diversity

Alfian Akbar Gozali and Shigeru Fujimura

Abstract Years after being firstly introduced by Fraser and remodeled for modern application by Bremermann, genetic algorithm (GA) has a significant progression to solve many kinds of optimization problems. GA also thrives into many variations of models and approaches. Multi-population or island model GA (IMGA) is one of the commonly used GA models. IMGA is a multi-population GA model objected to getting a better result (aimed to get global optimum) by intrinsically preserve its diversity. Localization strategy of IMGA is a new approach which sees an island as a single living environment for its individuals. An island's characteristic must be different compared to other islands. Operator parameter configuration or even its core engine (algorithm) represents the nature of an island. These differences will incline into different evolution tracks which can be its speed or pattern. Localization strategy for IMGA uses three kinds of single GA core: standard GA, pseudo GA, and informed GA. Localization strategy implements migration protocol and the bias value to control the movement. The experiment results showed that localization strategy for IMGA succeeds to solve 3-SAT with an excellent performance. This brand new approach is also proven to have a high consistency and durability.

Keywords Genetic algorithms · Island model genetic algorithm · Localization strategy · 3-SAT

1 Introduction

Genetic algorithm (GA) has been firstly introduced its idea by Fraser [6] and remodeled for modern application by Bremermann [2] in late 1950. After that time, GA, which also classified as metaheuristic algorithm, is commonly used especially to solve the optimization problem, for example, VLSI [1], university timetabling [23],

A.A. Gozali (✉) · S. Fujimura (✉)
Graduate School of IPS, Waseda University, Kitakyushu, Fukuoka 808-0135, Japan
e-mail: alfian@fuji.waseda.jp

S. Fujimura
e-mail: fujimura@waseda.jp

© Springer International Publishing AG 2018
R. Lee (ed.), *Computer and Information Science*, Studies in Computational Intelligence 719, DOI 10.1007/978-3-319-60170-0_11

knapsack problem [12], and order picking problem [22]. There are many variations of GA such as GA [8], fGA [24], pseudo GA [4], AIMGA [9], informed GA [23], hyper GA [5], even hybrid GA [21]. Between these variations, AIMGA, fGA, and hyper GA are included as a multi-population approach or usually called as island model GA. The main advantages of island model GA over other variants are its scalability and better ability to escape from local optimum.

In island model GA, diversity of the populations intrinsically is preserved. As a result, populations which are sets of solutions can be enlarged their ranges and give an advantage in system efficiency. The common problem usually met in island model GA is the population tends to converge toward local optima. Premature convergence is a term to represent this problem which occurs too early. The reason behind this is genetic (allelic) drift which refers to the frequency change of a gene variant (allele) in a population due to a random sampling of organisms [14, 18]. In island model GA, it leads to loss of diversity produced by the usage of finite population sizes.

Premature convergence in island model GA is a consequence of migration mechanism. The idea behind this is to migrate several individuals (usually the best one) from an island into another island to keep the diversity of the system. But in implementation, migration mechanism in a certain time will cause genetic drift. Another reason is because the genetic operator configurations (mutation probability, crossover probability, and population size) are same.

In a real-world nature, the reason for individuals migration is to find a more potentially living environment [7, 20, 25]. They migrate from a place with certain characteristics to another place with different characteristics which is potentially better. Ray [17] stated that living environment (medium) is one of the challenges to inoculate natural evolution into artificial media. This is because the evolutionary process is mainly concerned with adaptation to the living environment.

The islands conditioning to become a living environment with different characteristics (configurations) can be a solution to solve genetic drift problem. Different configuration can make different evolution speed and solution range. This will force each island to generate different characteristic of an individual that can lead to different local optimum. So, this mechanism will preserve the diversity between islands.

Localized living environment is a nature-inspired mechanism which has high potential to be implemented in island model GA to preserve the diversity of populations. Two main research questions investigated in this chapter are how to implement localization strategy for island model GA and how is its performance in solving a common optimization problem.

2   Localized Island Model GA

Island model GA (IMGA) is a multi-population GA model objected to getting a better result (aimed to get global optimum) by intrinsically preserve its diversity. Population in IMGA is usually called island after real-world definition that an island is always separated each other and has a population. An island is usually seen as a group of individuals (population) which represents a set of solutions. Almost there is no difference between one island and the others except its population.

Localization strategy of IMGA is a new approach which sees an island as a single living environment for its individuals. Because of that, an island's characteristic may be different compared with other islands. The characteristic of an island is represented as its operator parameter configuration or even its core engine (algorithm). The configuration differences will incline into different evolution tracks which can be its speed or pattern. An island can evolve faster or go into different evolution pattern of individual rather than the other. Taken together, the different evolution mechanism of each island can lead to a different pattern of an individual in a population. Or in another word, the diversity of islands will be well preserved.

Currently, there is a similar approach to get a better result by differencing GA mechanism. The latest research is implemented in a single system named evolutionary algorithm hyper-heuristics (EAHH) [13, 16, 19]. This algorithm tries to combine evolutionary algorithm (EA) with (mostly) constructive hyper-heuristics. EAHH uses several low-level heuristics to search the best heuristic to generate solutions. Despite its great performance, EAHH has drawbacks such as it takes much time in selecting heuristic rather than generating a solution, there must be more than one low-level heuristic, and it has relatively high complexity running on a single computer.

The distributive mechanism of Localization strategy for IMGA hopefully can fix a general EAHH problem. This is because in IMGA even one heuristic is enough, GA complexity will be separated into different computers, and it focusing on searching the best solution. For compensation, the diversity will be preserved with a new localization strategy introducing in this research.

2.1 Localization Strategy

To determine a correct strategy to localize islands in IMGA, there are several things have to be considered. First, it is how we differ the living environment between an island and the others. As mentioned before, the different living environment could mean different operator parameter configuration or its GA core engine (algorithm). The differences must incline into specific track (goal) differences. So, we classified GA into two big approaches: speed- and performance-based approach.

Speed-based approach is GA variant which tends to patch up its computational speed. This kind of GA tries to get a good result as fast as possible. The examples of this variant are pseudo GA (PGA) [4] and AIMGA [9]. Contrary, performance-based approach tries to maintain its populations' diversity balance so the global optimum can be reached even though it takes more time. The examples of this variant are fGA [24], hyper GA [5], and hybrid GA [21]. Furthermore, there is also a performance-based variant which suffers its computational time to get a better result by implementing local search and greedy operation such as informed GA [23].

These two kinds of GA variants imply that localized islands in IMGA at least should consist of a standard GA (SGA) island, a speed-based island, and a performance-based island. This research will use islands configuration such as standard GA (SGA), pseudo GA (PGA), and informed GA (IGA). PGA and IGA were chosen because these two GA models are single model GA. Single model GA means a type of GA model which runs on a computer so it will well fit with the concept of an island. Each of algorithm structures of SGA, PGA, and IGA is shown in Fig. 1.

Each of algorithms takes a role as an island. Besides those three islands, there is also an island called *master island*. Master island takes a role in controlling migration between those islands which are usually named as *slave island*. The interaction between slave and master islands is shown in Fig. 2. There are three migrant win-

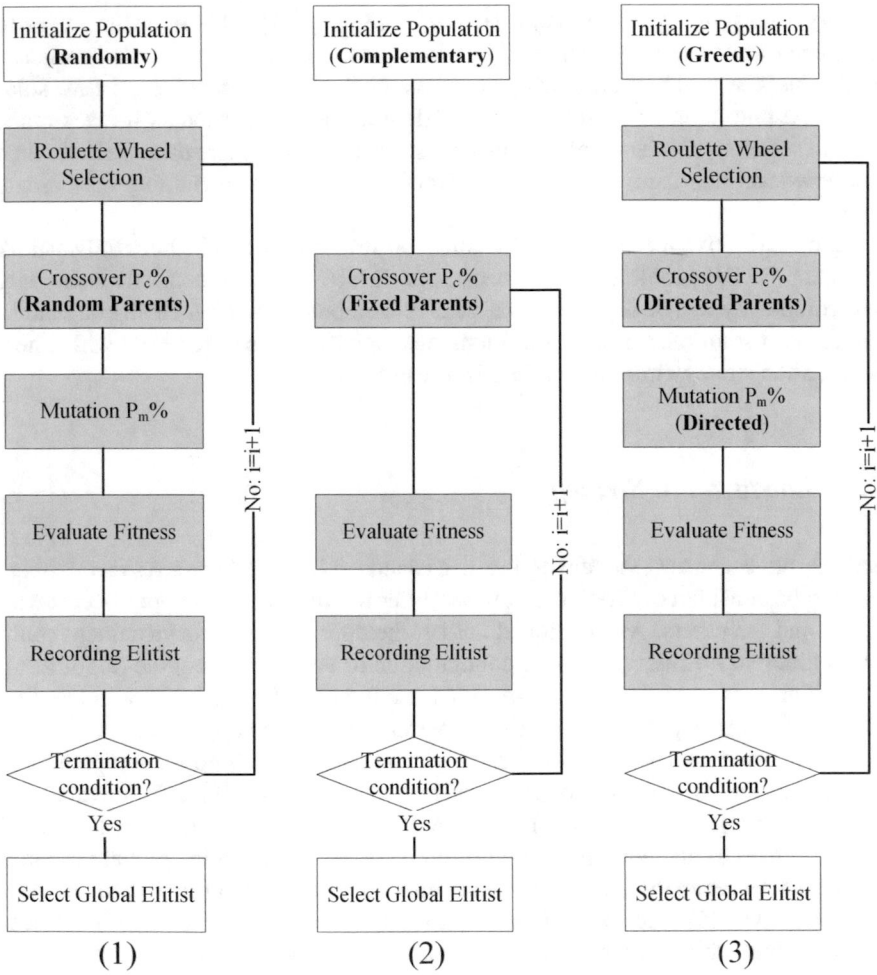

Fig. 1 Algorithm structures of SGA(1), PGA(2), and IGA(3)

Fig. 2 Interaction between
master and slave islands

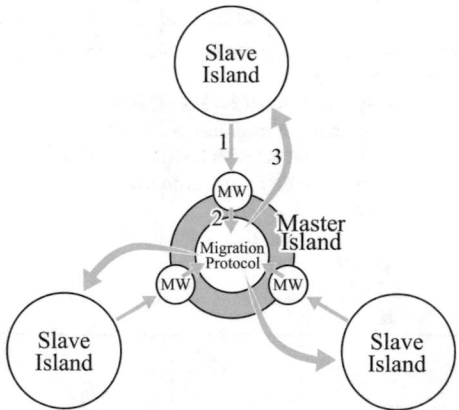

dows (MW) in master island to keep the best individual from the last generation of
every island. Because each slave islands has very different GA model, the migration
mechanism must be asynchronous as implemented in AIMGA [9]. The asynchronous
mechanism will make each island isolate but still have communication. A bit different from AIMGA which used fixed sequential migration mechanism, the proposed
IMGA model will use *migration protocol*.

Firstly, slave island completed its generation put the best individuals and current
bias value (B^t) in its migrant window. Next, migration protocol determines whether
this is the time for migration or not. If this is yes, master island sends the selected
individual to particular island determined by migration protocol algorithm.

2.2 Migration Protocol

Because IMGA implements asynchronous migration mechanism, it must be ensured
that the migration will not interrupt the island's core GA computational process.
There will be a migrant window in master island to keep the best individual for each
island's last generation. This individual will migrate if there is an island which is
indicated to tend into premature convergence. Therefore, IMGA needs a migration
protocol to rule and control its migration mechanism. The pseudocode of migration
protocol is explained in Algorithm 1. Let θ be the threshold constant and B_P^t is bias
value of island P in generation t.

The migration protocol adapts bias value from fGA [24]. Bias value will be used to
evaluate premature convergence tendency of an island. A population P^t at generation
t in an island can be represented as following matrix $N \times L$. Where N is the number
of individuals on that island, and L is the length of individual string. Each row vector
represents the string of an individual and $P_{i,j}^t$ is its allele.

Algorithm 1 Migration Protocol Pseudocode

Require: an island P send the best individual (p_i^t)
  if $B_P^t \geq \theta$ **then**
    if migrant Windows $\neq$ *null* **then**
      if number of migrants > 1 **then**
        find migrant with furthest HD from p_s^t
        migrate that migrant to island P
      else
        migrate migrant to island P
      end if
    end if
  end if

$$P(t) = \begin{bmatrix} p_{1,1}^t & p_{1,2}^t & \cdots & p_{1,L}^t \\ p_{2,1}^t & p_{2,2}^t & \cdots & p_{2,L}^t \\ \vdots & \vdots & \ddots & \vdots \\ p_{N,1}^t & p_{N,2}^t & \cdots & p_{N,L}^t \end{bmatrix} \tag{1}$$

The bias value B^t is defined as the measure of genotypic diversity of population $P(t)$ and $0.5 \leq B^t \leq 1.0$. The bias formula is shown in Eq. 2.

$$B_P^t = \frac{1}{N \times L} \sum_{j=1}^{L} \left(\left| \sum_{i=1}^{N} p_{i,j}^t - \frac{N}{2} \right| + \frac{N}{2} \right) \tag{2}$$

B^t is the convergence indicator which shows average percentage of the most prominent value in each position on the individuals. Larger values mean high convergence or low genotype diversity. By implementing bias value, the migrant window must be changed to be not only just keeping the best individual but also bias value from the island's last generation.

Derived from slave–master interaction in Fig. 2 and migration protocol in Algorithm 1, the asynchronous migration mechanism for master and slave islands can be represented as state diagram in Figs. 3 and 4.

After it was initialized, the master island has *idle* state until a slave sends its best individual. Master is in *keep* state while comparing that slave's bias value with a predefined threshold. If it is more than or equals with the threshold and there is any migrant except that slave's migrant window then *compute hamming distance* for each of migrants in a migrant window. Next, migrant with the furthest distance will be *migrated* to a slave which currently sent the individual. In another hand, slave island is in *compute* state after initialization. It computes core GA process as shown in Fig. 1 which colored in gray. Right after finish its generation, slave *sends* its best individual to master island. If the population in slave island tends to converge which is represented by its bias value, it will receive *migrated* individual from the master island.

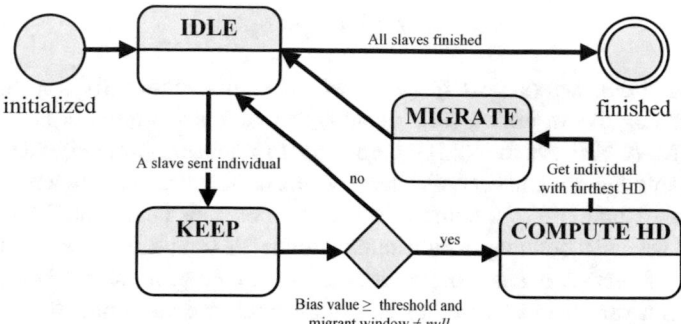

Fig. 3 Master island state diagram

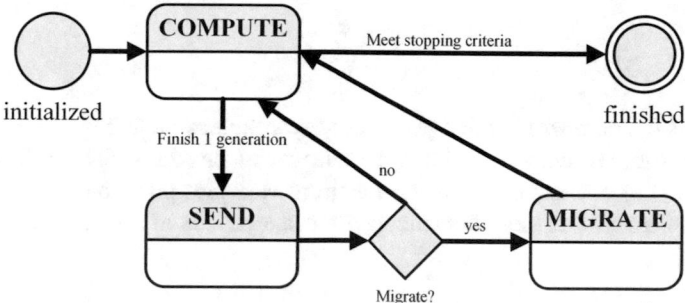

Fig. 4 Slave island state diagram

3 Implementation for Island Model GA

As mentioned in Sect. 2.1, localization strategy which is used in this research is IMGA with 3 localized slave islands. These islands consist of SGA, PGA, and IGA. In order to control migration between these islands, master island and migration protocol mechanism in Sect. 2.2 will be used. For the problem case, we used 3 Boolean satisfiability problem (3-SAT). The consideration picking this case is because 3-SAT is one of the essential NP-complete problems [3]. 3-SAT is simple but it can be very complex in accordance by its scaling. Details of the implementation are as follows.

3.1 Fitness Function

Boolean satisfiability problem (SAT) is a problem trying to get true result for a clause by setting true or false for every Boolean variables in a clause [11]. 3-SAT problem is SAT with 3 variables per clause. Let a propositional formula as:

$$(x_1 \lor x_2 \lor x_3) \land (x_1 \lor x_2 \lor x_3) \tag{3}$$

Formula 3 consists of a set of variables and logical connectives restricting the values of using the variables, such as $\neg$, $\lor$, and $\land$. This formula is in conjunctive normal form (CNF) which is a standard form to represent propositional formulas. CNF contains the conjunction of a number of clauses, the place where a clause is definitely a disjunction of a number of variables or their negations. Most previous researches used clauses conjunction summation as its fitness function as in [11, 15]. That fitness function is easy to understand, but it cannot represent how much the effect of each variable is. Therefore, in this research, we use a new fitness function based on the variables as follows.

$$f = \sum_{i=1}^{n} V_i \tag{4}$$

Let V_i is the number violation per clause of variable x_i including its negation $-x_i$, m is the number of clauses, and n is the number of variables. Where $(0 \le V_i \le m)$ and $(1 \le i \le n)$. By using this fitness function, variables those have more numbers than the others will be more influencing. Our objective is to minimize the f value.

3.2 Chromosome Representation

This research uses binary chromosome representation. The chromosome consists of genes which represent the value of variables $(1 = true, 0 = false)$.

3.3 Standard Genetic Algorithm

The GA core parameters configuration of the slave islands mainly uses reference from [4] for population size (μ), crossover probability (P_c), and mutation probability (P_m). After that reference, this research uses $\mu = 20 - 40$, $P_c = 50\% - 75\%$, and $P_m = 2\%$. SGA has a common traditional GA process with roulette wheel selection. It is shown in Fig. 1.

3.4 Pseudo Genetic Algorithm

Parameter configuration for PGA is same with SGA but with different processes. PGA implements no roulette wheel selection and mutation. For the consequences,

PGA will use static–dynamic complementary chromosome [4] for initialization and crossover. It is a mechanism in PGA to avoid incest breeding by generating the complement of a parent chromosome in order to be its couple.

3.5 Informed Genetic Algorithm

IGA processes are almost similar with SGA. The differences are in the detail of processes, especially for initialization and mutation phase. IGA uses greedy initialization and directed mutation.

Steps of greedy initialization (execute sequentially):

1. Find all variables which do not have negation. Remove all clauses containing them.
2. Sort variables order by its positive–negative ratio then for the top 20% variables, set their value to 1 if positive > negative and 0 if negative < positive.

For directed mutation, we use three steps mutation operators. Directed mutation is performed to get better fitness faster. These steps are executed sequentially. Although the first mutation produced a better chromosome, the next mutation is still executed to get further better chromosomes. These are the three steps of directed mutations:

1. Find all violated variables and negate them start from the bigger violation numbers. Keep if the fitness increases, revert if decreases.
2. From all violated variables, pick two variables which have different value randomly then swap their value.
3. From all violated variables, pick two variables which do not share same clause connection then complement their value.

4 Experiment Results

This research uses SATLIB [10] benchmark problems to analyze localization strategy of IMGA. SATLIB is widely known as a database for SAT benchmark, especially for 3-SAT. This research uses uniform random 3-SAT. Total ratio of the clauses and variables in these problems is 4.3.

The experiment goals for this research are three. The first goal is ensuring localization strategy is well implemented for IMGA proven with SATLIB dataset. The second goal is analyzing the performance of localization strategy compared with current research in IMGA for 3-SAT conducted by Salmah et al. [15]. For comparison need, we use the exactly same suit with [15] as follows in Table 1.

The last experiment goal is analyzing the robustness of localization strategy for IMGA when solving a large problem set. For the last experiment, we use a full suit

Table 1 The SAT problem in the experiment

Suit	Instances	Variables (n)	Clauses (m)
URSAT1	10	20	91
URSAT2	10	50	218
URSAT3	10	75	325
URSAT4	10	100	430

Table 2 The Full SAT problem in the experiment

Suit	Instances	Variables (n)	Clauses (m)
uf20-91	10	20	91
uf50-218	10	50	218
uf75-325	10	75	325
uf100-430	10	100	430
uf125-538	10	125	538
uf150-645	10	150	645
uf175-753	10	175	753
uf200-860	10	200	860
uf225-960	10	225	960
uf250-1065	10	250	1065

Table 3 The fitness comparison between SGA, Salmah's IMGA, and localized IMGA

Suit	SGA	Salmah's IMGA	Localized IMGA
URSAT1	91.00	91.00	91.00
URSAT2	212.01	217.195	218.00
URSAT3	306.67	323.85	324.38
URSAT4	402.38	428.23	429.85

of uniform random 3-SAT from SATLIB. Detail of the benchmark dataset is shown in Table 2.

For evaluation parameter, the experiments use fitness value, a number of the true clauses, and average run time. Because previous research used a different kind of fitness, for comparison purpose, the first and second experiments use number of the true clause and average run time. The fitness value is used for the third experiment to analyze how localization strategy works in IMGA.

Table 3 shows the comparison between SGA, IMGA by Salmah et.al., and IMGA with localization strategy. The result shows that for the first suit (URSAT1), all GA models can reach maximum fitness 91 which means every clause is satisfiable. The second suit (URSAT2) localized strategy for IMGA can reach maximum fitness for every instance. And for the last 2 suits, there is no GA model which can reach maxi-

Table 4 The durability test result

Suit	Best fitness	Number of best fitness	Average fitness
uf20-91	0	10	0
uf50-218	0	10	0
uf75-325	0	5	3.2
uf100-430	0	4	4.4
uf125-538	0	6	4.7
uf150-645	0	5	8.5
uf175-753	0	5	10.7
uf200-860	0	4	14.9
uf225-960	0	3	17.8
uf250-1065	0	4	18.3

mum fitness, but IMGA with localization strategy can get the best result compared to the other two. From Table 3, it can also be seen that localization strategy for IMGA can significantly improve the effectiveness of IMGA compared with the previous model (Salmah's IMGA).

For the durability test of localization strategy for IMGA, new fitness formulation was used. Because of minimization objective function, lower fitness means better result and the optimum result is 0. For durability test, localization strategy for IMGA will be run 10 times every suit. The average fitness and the best fitness are recorded. The result of the last test is shown in Table 4. From that result, localization strategy for IMGA has high durability because it still can produce the best (zero) fitness for every test suit. It also has a high consistency proven by the number of zero fitness value which occur more than 3 times for every test suit.

5 Conclusion

Localization strategy for IMGA is a brand new approach in multi-population GA model which try to differ the environment of its populations. Localization strategy for IMGA is proven to solve one of the essential problem in optimization, Boolean satisfiability problem with three clauses (3-SAT). Compared with previous research using IMGA to solve this problem, localization strategy for IMGA produced significant improvement for same test suits. For the durability test, with more complex test suit, localization strategy for IMGA can reach the best fitness at least in 3 of 10 trials.

Taken together, localization strategy gives great improvement for IMGA. According to experiment results, localization strategy for IMGA has high potential to be implemented in more complex optimization problem such as job shop scheduling, university course timetabling problem. The localization strategy can still be improved further by implementing another heuristic in migration protocol or by using other variations of GA model. However, it still must be analyzed further for the network cost for every slave islands. Comparison with the other current solvers for optimization with the different approaches such as hyper-heuristics or evolutionary algorithm hyper-heuristics is needed to analyze its performance deeper.

Acknowledgements This work was supported by Indonesia Endowment Fund for Education (LPDP), a scholarship from Ministry of Finance, Republic of Indonesia. This work was conducted while at Graduate School of Information, Production, and Systems, Waseda University, Japan.

References

1. Bhuvaneswari, M.: Application of Evolutionary Algorithms for Multi-objective Optimization in VLSI and Embedded Systems. SpringerLink: Bcher. Springer India. https://books.google.co.jp/books?id=ybVTBAAAQBAJ (2014)
2. Bremermann, H.J.: The evolution of intelligence: the nervous system as a model of its environment. Techreport 1, Deparment of Mathematics, University of Washington, Seattle (1958). Contract No. 477(17)
3. Cheeseman, P., Kanefsky, B., Taylor, W.M.: Where the really hard problems are. In: Proceedings of the 12th International Joint Conference on Artificial Intelligence, IJCAI'91, vol. 1, pp. 331–337. Morgan Kaufmann Publishers Inc., San Francisco, CA, USA. http://dl.acm.org/citation.cfm?id=1631171.1631221 (1991)
4. Chen, Q., Zhong, Y., Zhang, X.: A pseudo genetic algorithm. Neural Comput. Appl. **19**(1), 77–83 (2010). doi:10.1007/s00521-009-0237-3
5. Cowling, P., Kendall, G., Han, L.: An investigation of a hyperheuristic genetic algorithm applied to a trainer scheduling problem. In: Proceedings of the 2002 Congress on Evolutionary Computation, CEC 2002, vol. 2, pp. 1185–1190 (2002). doi:10.1109/CEC.2002.1004411
6. Fraser, A.: Simulation of genetic systems by automatic digital computers. Aust. J. Biol. Sci. **10**(2), 492–499 (1957)
7. Garcia, A.J., Pindolia, D.K., Lopiano, K.K., Tatem, A.J.: Modeling internal migration flows in sub-saharan africa using census microdata. Migr. Stud. **3**(1), 89 (2014). doi:10.1093/migration/mnu036
8. Goldberg, D.E.: Genetic Algorithms in Search, Optimization and Machine Learning, 1st edn. Addison-Wesley Longman Publishing Co., Inc, Boston, MA, USA (1989)
9. Gozali, A.A., Tirtawangsa, J., Basuki, T.A.: Asynchronous island model genetic algorithm for university course timetabling. In: Proceedings of the 10th International Conference on the Practice and Theory of Automated Timetabling, pp. 179–187. PATAT, York (2014)
10. Hoos, H.H., Sttzle, T.: Satlib: an online resource for research on sat. In: SAT 2000, pp. 283–292. IOS Press (2000)
11. Li, J., Wang, H., Liu, J., Jiao, L.: Solving sat problem with a multiagent evolutionary algorithm. In: 2007 IEEE Congress on Evolutionary Computation, pp. 1416–1422 (2007). doi:10.1109/CEC.2007.4424637
12. Li, Q.J., Szeto, K.Y.: Efficiency of adaptive genetic algorithm with mutation matrix in the solution of the knapsack problem of increasing complexity. In: 2015 IEEE Congress on Evolutionary Computation (CEC), pp. 31–38 (2015). doi:10.1109/CEC.2015.7256871
13. Li, W., Zcan, E., John, R.: Multi-objective evolutionary algorithms and hyper-heuristics for wind farm layout optimisation. Renew. Energy **105**, 473–482 (2017). doi:10.1016/j.renene.2016.12.022. http://www.sciencedirect.com/science/article/pii/S0960148116310709
14. Masel, J.: Genetic drift. Curr. Biol. **21**(20), R837–R838 (2011). doi:10.1016/j.cub.2011.08.007. http://www.sciencedirect.com/science/article/pii/S0960982211008827

15. Mousbah Zeed Mohammed, S., Tajudin Khader, A., Azmi Al-Betar, M.: 3-SAT using island-based genetic algorithm. IEEJ Trans. Electron. Inf. Syst. **136**(12), 1694–1698 (2016). doi:10.1541/ieejeiss.136.1694. https://www.jstage.jst.go.jp/article/ieejeiss/136/12/136_1694/_article

16. Raghavjee, R., Pillay, N.: A genetic algorithm selection perturbative hyper-heuristic for solving the school timetabling problem. ORiON **31**(1), 39–60 (2015)

17. Ray, T.S.: An evolutionary approach to synthetic biology: Zen in the art of creating life. Artif. Life **1**(1), 179–209 (1993)

18. Rogers, A., Prugel-Bennett, A.: Genetic drift in genetic algorithm selection schemes. IEEE Trans. Evol. Comput. **3**(4), 298–303 (1999). doi:10.1109/4235.797972

19. Rosanne Els, N.P.: An evolutionary algorithm hyper-heuristic for producing feasible timetables for the curriculum based university course timetabling problem. In: 2010 Second World Congress on Nature and Biologically Inspired Computing. Kitakyushu (2010)

20. Schnell, P., Azzolini, D.: The academic achievements of immigrant youths in new destination countries: evidence from southern europe. Migr. Stud. **3**(2), 217 (2014). doi:10.1093/migration/mnu040

21. Segredo, E., Lalla-Ruiz, E., Hart, E., Paechter, B., Voss, S.: Hybridisation of Evolutionary Algorithms Through Hyper-heuristics for Global Continuous Optimisation, pp. 296–305. Springer International Publishing, Cham (2016)

22. Stauffer, M., Hanne, T., Dornberger, R.: Uniform and non-uniform pseudorandom number generators in a genetic algorithm applied to an order picking problem. In: 2016 IEEE Congress on Evolutionary Computation (CEC), pp. 143–151 (2016). doi:10.1109/CEC.2016.7743789

23. Suyanto, S.: An informed genetic algorithm for university course and student timetabling problems. In: Artificial Intelligence Soft Computing. Lecture Notes of Computer Science, vol. 6114, pp. 229–236. Springer, Berlin (2010)

24. Tsutsui, S., Fujimoto, Y., Ghosh, A.: Forking genetic algorithms: gas with search space division schemes. Evol. Comput. **5**(1), 61–80 (1997). doi:10.1162/evco.1997.5.1.61

25. Vardanis, Y., ke Nilsson, J., Klaassen, R.H., Strandberg, R., Alerstam, T.: Consistency in long-distance bird migration: contrasting patterns in time and space for two raptors. Anim. Behav. **113**, 177–187 (2016). doi:10.1016/j.anbehav.2015.12.014. http://www.sciencedirect.com/science/article/pii/S0003347215004558

HM-AprioriAll Algorithm Improvement Based on Hadoop Environment

Wentian Ji, Qingju Guo and Yanrui Lei

Abstract In order to improve the efficiency of the mining frequent item-sets of AprioriAll algorithm, the Hadoop environment and MapReduce model are introduced to improve AprioriAll algorithm, a new algorithm of mining frequent item-sets under the environment of big data HM-AprioriAll algorithm is designed. Compared with the original algorithm, the new algorithm introduces user attributes and pruning technology, which reduces the number of the elements in the candidate sets and reduces the number of the scanning times on the data sets, greatly reduces the time complexity and space complexity of computing, gives rules model in large scale. After testing HM-AprioriAll algorithm on Hadoop platform, the results prove that the insertion of this technology makes HM-AprioriAll algorithm have higher efficiency of expanding.

Keywords Hadoop · AprioriAll algorithm · Map Reduce · Recommendation systems · Big data

W. Ji
Department of Software Engineering, Hainan College of Software Technology,
Fuhai Road No. 128, Qionghai, People's Republic of China
e-mail: skywarps@163.com

Q. Guo (✉)
Department of Information Management, Hainan College of Software Technology,
Fuhai Road No. 128, Qionghai, People's Republic of China
e-mail: guoqj8859@163.com

Y. Lei
Department of Network Engineering, Hainan College of Software Technology,
Fuhai Road No. 128, Qionghai, People's Republic of China
e-mail: callouswater@hotmail.com

© Springer International Publishing AG 2018
R. Lee (ed.), *Computer and Information Science*, Studies in Computational
Intelligence 719, DOI 10.1007/978-3-319-60170-0_12

1 Introduction

With the rapid development of information technology and Internet, the Internet accommodates more and more information. With the "explosive" growth of e-commerce market, social networks, and mobile communication service, we are now in an era of "information overload" [1]. How to filter out the useless information and display the information which users are really "interested in" in the comprehensive information is the new problem and new challenge confronted by the development of Internet technology.

Data mining technology, with the environment of the integration of Internet, the Internet of Things and cloud computing and data explosion, also obtains the development which progresses by leaps and bounds. Data mining technology refers to the process which extracts useful, potential, and relevant knowledge from a large amount of data, and it is an important step of knowledge discovery KDD [2], in which the sequence pattern mining Apriori algorithm is one of the most classic algorithms.

In the study of association rules, one of the most famous algorithm is Apriori algorithm put forward by Rakesh Agrawal and Ramakrishnan Srikant in 1994, in addition, there is FP-growth method, MINWAL, and DWAR algorithm which are found in view of the weighted association rules, etc.

The basic idea of the algorithms is as follows:

The main idea of Apriori algorithm is a recursive iteration process. Firstly, to find a frequent item-sets L_1 which contain a project, then recursively calculate the frequent item-set L_2 which contains two projects, until mine out all frequent item-sets containing all the projects, make the candidate item-sets empty. In the process, connection and pruning are applied to realize:

Connection: By the connection between L_{k-1} with itself to generate item-set C_k of candidate set K;

Pruning: C_k is a super-set of L_k, compress C_k, can quicken the completion of the sequence of the frequent itemsets.

The algorithm requires to scan database repeatedly and match to get a large set of candidate sets. On the basis of the principle of this algorithm, to improve the effectiveness of the algorithm, at present many scholars have proposed many improvements on Apriori algorithm, including using hash technology, things compression technology, classification counting, data set sampling technology, dynamic item-sets technology and iceberg querying, and other technologies, which have greatly improved the effectiveness of the algorithm.

But with the increase of the amount of data, the improvement of Apriori algorithm on association rules is also increasing, which mainly reflects in how to reduce the candidate sets and processing time. This paper puts forward the AprioriAll algorithm improvement based on the framework of MapReduce parallel computing model under Hadoop environment and tests the improved algorithm on Hadoop platform, and the results show that the computing efficiency has improved significantly.

2 Sequential Pattern Mining Algorithm

2.1 The Theory of Sequence Pattern Mining

Sequential pattern mining is an important research branch of data mining, and it is mainly a knowledge discovery process which aims to find frequent subsequence from sequence data sets and make it as a model [2]. For its characteristics of strong practicability and easily being understood, it has great influence in the field of data mining research. From sequential pattern mining algorithm Apriori firstly proposed in 1994 to now, a lot of classic algorithms for mining sequential patterns are proposed, including a lot of improved algorithms, such as: AprioriAll, Dynamic-Some, generalized sequence mining, GSP algorithm [3], FreeSpan algorithm based on vertical data format, and so on, which have improved the efficiency of data mining to a certain extent. But with the advent of the era of big data, in the face of the growing data scale and a variety of complex structures in different forms, there occurs various problems such as memory overflow, redundancy item-sets, large overhead of time, and space when using traditional sequential patterns to do data mining, which greatly reduces the performance of the traditional sequential patterns mining algorithm and cannot meet the real demands of customers in the era of big data.

2.2 The Basic Concepts of Sequential Pattern

Definition 1 Sequence database D, $D = \langle t_1, t_2, \ldots, t_n \rangle$, in which t_i is for each transactions.

Definition 2 Item-sets I, $I = \langle I_1, I_2, \ldots, I_n \rangle$ and composed by a different set of items (Item).

Definition 3 Sequence S, $s = \langle s_1 s_2 \ldots s_n \rangle$ different item-sets are composed in a certain order, and among them, S_i is for sequence S elements.

Definition 4 Subsequence: Set sequence $s' = \langle s_1' s_2' \ldots s_m' \rangle$, $s = \langle s_1 s_2 \ldots s_n \rangle$ and when $1 \leq i_1 < i_2 < \ldots i_m \leq n$, $s_1' \subseteq s_{i1} \ s_2' \subseteq s_{i2} \ s_m' \subseteq s_{im}$ known as s' is the subsequence of S, record as $s' \subseteq s$. The Association Function of Large-scale Network Survivability.

Definition 5 The support: refers to the number of the sequences which contain se-quence S' in sequence database D, expresses as the support (S').

Definition 6 Frequent Sequence: for the minimum support threshold, if the support of the sequence in data set is greater than or equal to the threshold, then the sequence is called Frequent Sequence.

2.3   The Process of the Sequence Pattern Mining

According to the definition in B, the sequence pattern mining is mainly divided into three stages including data preprocess, candidate model set, and sequences model set. The sequential pattern mining stages are shown in Fig. 1:
   It is mainly divided into the following three steps:

(1) Data preprocess: According to the data source, to make data preprocess on data mainly includes to sort and classify the sequence data set according to certain rules and form a sequence.
(2) In Sequence I, find the sequences with the length of $K - 1$, iterate until the new frequent item-set cannot be generated, namely, its support must meet the minimum support.
(3) According to the rules and the training of the knowledge base, find a complete set of patterns from the candidate model, until the sequential pattern is generated.

At present, the research of sequential pattern mining algorithm from domestic and foreign areas is mainly divided into the algorithm based on Apriori algorithm and the algorithm based on pattern growth strategy, the former needs candidate sequences to make iterative calculation to produce corresponding sets of frequent sequential patterns, the latter needs the concepts of prefix and projection and does not need to generate candidate frequent patterns. Both have their own advantages. This paper mainly explores and studies on the improvement of AprioriAll algorithm based on Apriori characteristics.

2.4   AprioriAll Algorithm

Researchers divide the frequent sequences according to the generation of the sequences in horizontal and vertical way and propose many sequential pattern mining algorithms with the characteristics of Apriori [4]; at present there are also a

Fig. 1 Process of sequence pattern mining

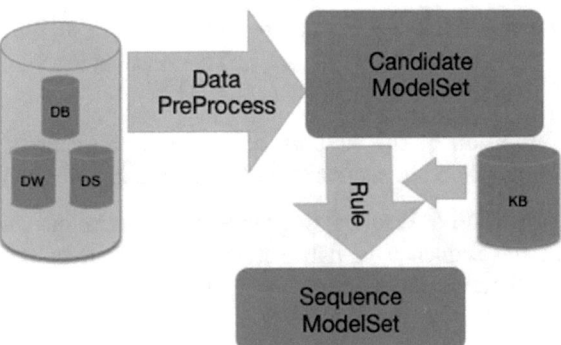

lot of improved algorithms, and AprioriAll algorithm is a kind of algorithm using level data format [5]. This algorithm constantly scans the database and adopts continuously "candidate-screening" iterative way to get all the frequent sequential pattern sets. Comparing with Apriori algorithm, the biggest difference is whether making full cross or forward intersection of former C_{k-1} elements when generating candidate set of C_k every time. In AprioriAll algorithm, the frequent item-set L_1 can be gotten in the first time of the scan log, then iterative method is used again to scan in the previous frequent items until no new strength sets occur, then the algorithm ends. Through the contrast analysis of the existing algorithms, we learn that the biggest problem of algorithm improvement is how to efficiently find rules and balance the complexity of the space and time, and improve efficiency. Therefore, the improvement goals of AprioriAll algorithm are naturally how to reduce the size of the candidate sets generated in each time and eventually to reduce the number of scanning. Because by several experimental observations, we find that in the huge candidate sets generated each time, the subsets which constitute any elements of candidate set Lk must appear in the last candidate set L_{k-1}, the real frequency should be few. There are many reasons, such as the algorithm does not take the association between users and transactions into account and does not properly make pruning on the candidate sets, thus the computational complexity of the algorithm will be increased.

3 HM-AprioriAll Algorithm Improvement Based on Hadoop Environment

In order to meet the growing huge amounts of data and the demand of the diversification mode, the traditional single machine, the processing mode under serial data environment has been unable to adapt to it. Cloud computing with its mass distributed storage and collaboration technology has become a new computing model in the era of big data. The Hadoop is a free and open source platform for cloud computing application, and users can apply the mass storage capacity of the cloud computing platform and MapReduce distributed programming model framework to realize the distributed tasks [6] even under the premise of knowing nothing about the distributed low-level details.

3.1 Key Technology

Hadoop is a reliable, efficient, and strong extensibility, open source distributed software framework [7] with the advantages of high processing capacity, low cost, cross-platform operation, and easy to be used. In addition to its own file system HDFS (Hadoop Distributed File System), it also provides different interfaces for different file systems (such as FTP, KFS, and S3,) to make interoperation.

Because HDFS has high fault tolerance and uses the code migration mechanism, it is very suitable for it to have an application processing with large data sets.

MapReduce technology can help more new programmers make parallel application program development without having to learn more about the underlying design, and its name mainly comes from two important operations in the functional programming model: map and reduce operation and the calculation process is roughly divided into input, map, reduce, and output phases, which are shown in Fig. 2.

In Hadoop, machines used to perform the task of MapReduce have two roles: Job Tracker is mainly used for the dispatching work, TaskTracker is used to perform the work, and there is only one set of Job Tracker in a Hadoop cluster. It is shown in Fig. 3.

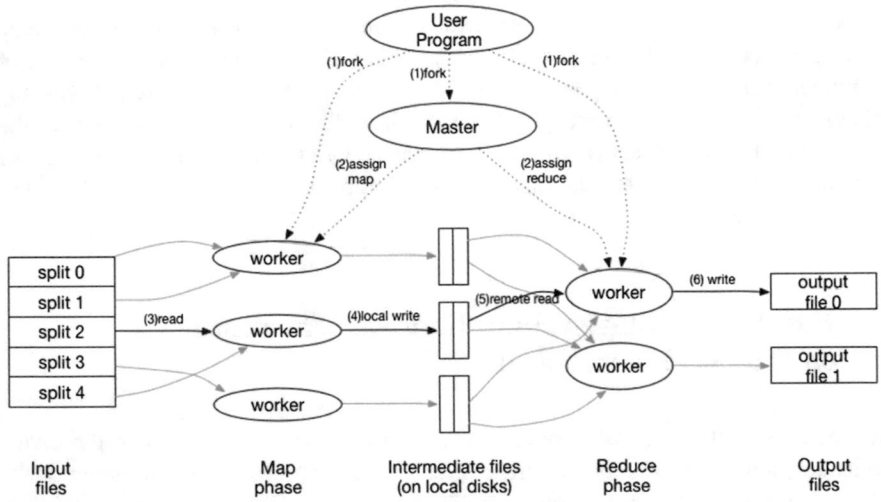

Fig. 2 Computing process of MapReduce

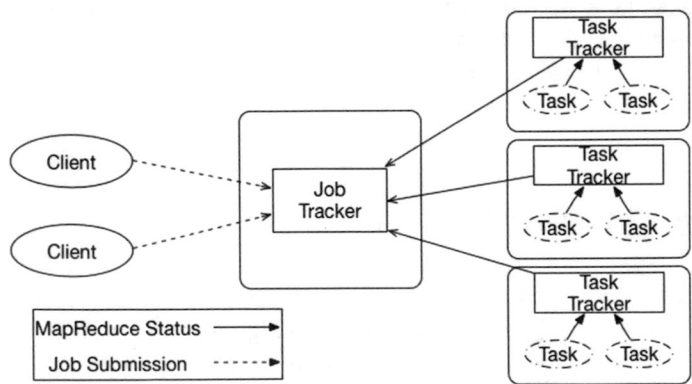

Fig. 3 Operating process of MapReduce

From Figs. 2 and 3, it can be found that HDFS and MapReduce commonly compose Hadoop distributed system, in which HDFS provides file operations and storage functions for the task processing of MapReduce, on the contrary, MapReduce has realized the task assignment, tracking, execution, and other processes and has collected the results on HDFS. Both interact with each other and highlight the two big main functions of Hadoop distributed cluster on task processing and distributed computing.

3.2 Algorithm Improvement

By experimental observations for many times, we find, in the huge candidate sets generated each times, the subsets which constitute any elements of candidate set L_k must appear in the last candidate set L_{k-1}, the real frequency should be few. There are many reasons, such as the algorithm does not take into account the association between users and transactions and does not properly make pruning on the candidate sets, thus the computational complexity of the algorithm will be increased.

To solve the above problems, we put forward two improved HM-AprioriAll algorithms based on the following steps, and the specific steps are as follows:

The first step: When generating candidate set Li each times, according to the support of the calculation in advance, delete the records which are less than the minimum support in the item-set and get frequent item-sets L_i, put forward to increase user attributes for each users, namely, User ID. Therefore, when making cross-generation of frequent item-sets, only users with the same User ID can process, otherwise the cross cannot be carried on, so that the repeatedly scanning will not be needed any more.

The second step: In the MapReduce model, the item-sets value are entered into the map function with the form of <Item, Value>, and output by map function, the reduce function code data with same value to the same reduce function, and complete the connection operation fast.

The third step: Do pruning operation for the candidate sets generated each time, put forward the algorithm of deleting the subset of the elements of candidate sets L_k whose elements are not in the last candidate sets L_{k-1}, and decrease the number of elements which have nothing to do with the set in the candidate sets. When new candidate sets do not generate any more, then the computing ends.

Through the above three steps of the two improvements, the elements of the candidate set reduce, which makes the scanning number of data sets reduce, at the same time it also reduces the time complexity and space complexity of the computing. The pseudocode of the improved algorithm is as follows:

Input:

U(UserSet) $= \{U_1, U_2, \cdots, U_n\}$

D(with UserID Session Database)=$\{S_{u1}, S_{u2}, \cdots, S_{uk}\}$

S (MinSupport)

Output: frequent item-sets

IF (UserID AND datetime of first page reference in each session)

D=Sort(D) ;

Find Lu,1 with UserID in D;

$L_{u,n}$=Improved_AprioriAll($U, D_u, S, L_{u,1}$);

Find maximal reference sequences from $L_{u,n}$;

Improved_AprioriAll($U, D_u, S, L_{u,1}$); n=2;

Do While $C_{u,n-1} <> null$

  Sort $L_{u,n-1}$ on UserID;

   For(i=1; i= Length($L_{u,n-1}$); i++)

    For(j=2;j=Length($L_{u,n-1}$); j++)

     IF(($i \langle \rangle j$) And ($U_i \cap U_j \langle \rangle Null$))

      $C_{ui \cap uj,n} = L_{u,n-1,i} \oplus L_{u,n-1,j}$;

       Append $C_{ui \cap uj,n}$ to $C_{u,n}$;

      End IF
      End For

     End For

  End While

For Each $C_{u,n}$ in $C_{u,n}$

   IF (not ($C_{u,n}$ SubSets) IN $L_{u,n-1}$)

   Delete $C_{u,n}$;

    IF Supprot($C_{u,n}$) IN D_u>=S

     Append $C_{u,n}$ to $L_{u,n}$;

     End IF

   End IF

    n=n+1;

   End For

End IF

4   The Experiment and Analysis

4.1   Experimental Environment

4.1.1   The Hardware Environment

The experiment adopts 4 PC computer clusters, with the configuration of Intel Core2.8 GHz, hard disk of 500 G, and memory of 4 GB DDR3, one is for master primary server, three are for slave support server, to achieve LAN interconnection between nodes.

4.1.2   The Software Environment

OS: Linux Ubuntu 12.04
Database: MySQL 5.5.41
Java 7, Tomcat 8.0, and Hadoop 1.0.4

4.1.3   Data Sources

The experiment adopts MovieLens which is often used in data mining experiment as the data set, and it mainly uses the users' opinion of movies, Web logs on the Web site, with a total of 1.88 GB, makes comparison test on the algorithm before and after the improvement, and analyzes the superiority of improving HM-AprioriAll algorithm.

4.2   Analysis of Experimental Results

There are 40 different items in 3021800 transactions of the experimental data, and the longest transaction has 33 items. After selecting 20% records of them and making 30 times of experiments, taking the average, the average time of frequent item-sets of getting access to can be gotten. Table 1 compares the algorithms before and after the improvement.

Figure 4 shows the comparison of the time consumption of mining K frequent item-sets of AprioriAll algorithm and HM-AprioriAll algorithm which is parallel with MapReduce based on the Hadoop environment under the condition with the increase of the number of frequent item-sets. It can be seen by the figure that when dealing with huge amounts of data, with the increase of mining frequent item-sets, the number of candidate sets exponentially level increases. When using MapReduce

Table 1 Comparison of AprioriAll algorithm before and after the improvement

Data size	Algorithm name	Average time
28 M	AprioriALL	20 min and 35 s
	HM-AprioriALL	9 min and 12 s

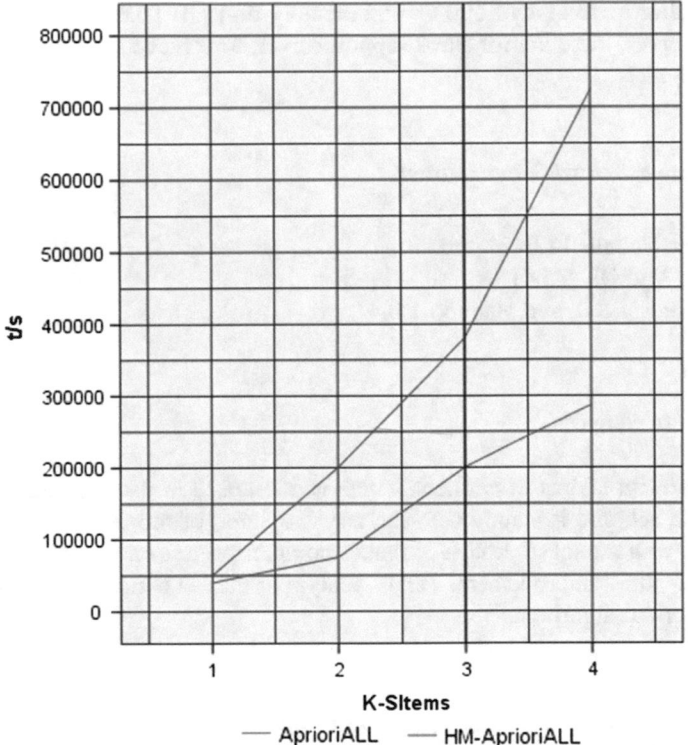

Fig. 4 Comparison of the time consumption of K frequent item-sets before and after the improvement of the algorithm

to parallel process and scan, if we increase the ID feature of the users and decrease the time of scanning, the operation efficiency will significantly improve.

Figure 5 shows the operational results of different number of nodes before and after the improvement of the algorithm; from the figure it can be seen that the more the nodes, the shorter the time it is needed for calculation, and the reason is that with the help of the distributed and parallel processing of MapReduce, the superiority of the proposed algorithm has greatly improved.

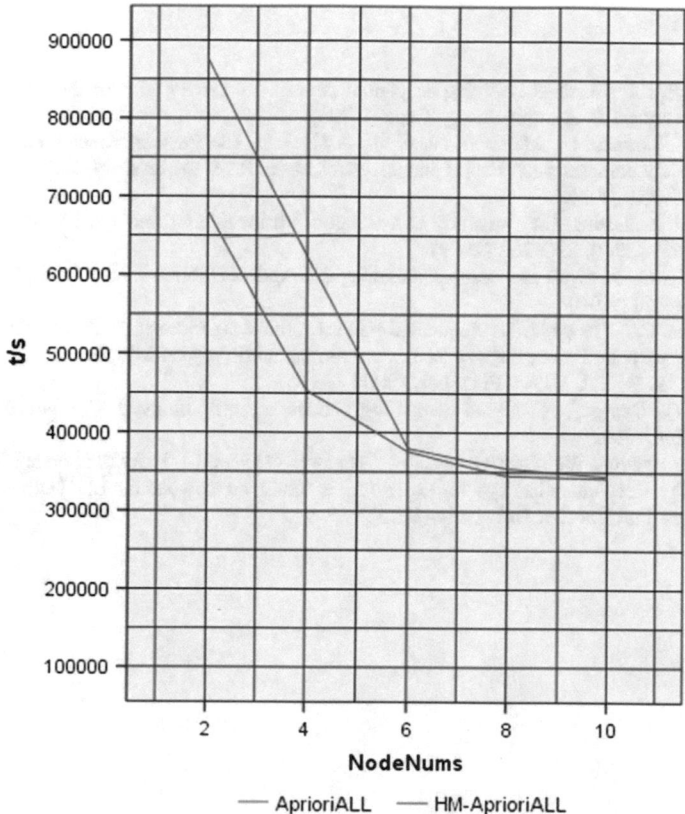

Fig. 5 Comparison of the operation time on different nodes before and after the improvement of the algorithm

5 Conclusion

This paper studies on the improvement of AprioriAll algorithm based on the Hadoop environment. On the basis of the existing research results, it introduces the HM-AprioriAll parallel cluster algorithm of MapReduce. Through experiment verification, the pattern matching of the candidate set in the case of reducing the number of database scanning has greatly improved the operation efficiency. In the era of big data, HM-AprioriAll algorithm has efficient operation ability for the mining of Web behavior pattern and provides a strong support for the mining of the users' implicit behaviors, and has good research value and application prospect.

Acknowledgements This work is supported by the college scientific research and fund project of Hainan province (Grant No. Hnky2016-67).

References

1. Agrawal, R., Srikant, R.: Mining sequential pattern. In: Proceedings of the 11 International Conference on Data Engineering, Taipei (1995)
2. Lara, J., Lizcano, D., Martinez, A., et al.: A UML profile for then conceptual modeling of structurally complex data: Easing human effort in the KDD process. Inf. Softw. Technol. **56** (3), 335–351 (2014)
3. Liqiang, Z., Dating, L.: The study of fast algorithm for mining frequent item-sets. J. Harbin Eng. Univ. **29**(3), 226–232 (2008)
4. Xiaoting, D.: Analysis and comparison of typical Apriori algorithm. Microcomput. Inf. **26**(3–3), 159–160 (2010)
5. Antunes, C., Oliveria, A.L.: Sequential pattern mining algorithms: trade-offs between speed and memory. In: Proceedings of 2nd International Workshop on Minging Graphs, Trees and Sequences (MGTS 2004), Pisa, Italy, 2004
6. Chen, Q., Deng, Q.-N.: Cloud computing and its key technology. J. Comput. Appl. **29**(9), 2562–2567 (2009)
7. Apache Hadoop: Welcome to Apache™ Hadoop@!.[EB/OL]. http://hadoop.apache.org/
8. Guo, Q.: Recommended system algorithm is study on application of elective course In: Colleges, Sun Yat-Sen University (2013)

Architecture of a Real-time Weather Monitoring System in a Space-time Environment Using Wireless Sensor Networks

Walid Fantazi and Tahar Ezzedine

Abstract This paper presents a Web-mapping framework for collecting, storing, and analyzing meteorological data recorded by wireless sensor network. The main objective is to design and implement climate monitoring system based on WSN able to intercepting and filtering meteorological records and generate alert system in real time in case of emergency. The application consists in transferring the data recorded by the sensor network via a gateway to an application server by using a CloudMQTT transfer protocol. The server, which allows intercepting sensor data in real time via Web Socket to a Web application (RIA), is designed to indexing and recording data in memory database. These data are stored in a spatial-temporal database and can be visualized via Web services API (REST, GWS).

1 Introduction

Today, wireless sensor networks have become a discipline that plays a very important role in the acquisition, use, and manipulation of information [1]. They are a less costly technique for observing the earth's surface that provides a wealth of data for different domains of environment [2], telecommunications [3] and transportation (followed by a road network) [4]. The majority of work based on wireless sensor networks is concerned with physical and network level study (routing, energy consumption, communication technique) and other solutions aiming at manipulating and analyzing the data recorded by sensor networks. Several works have been realized around architecture of monitoring application based on network sensors, in [5]. The authors proposed an approach for integrating wireless sensor nodes into a service-oriented architecture using a gateway to access the WSN. They suggested the use of service-oriented architecture (SOA) for integration into different systems such as

W. Fantazi (✉) · T. Ezzedine (✉)
Communication System Laboratory SysCom, National Engineering School of Tunis,
University Tunis El Manar, BP 37, 1002 Belvedere, Tunis, Tunisia
e-mail: Fantazi_w@yahoo.fr

T. Ezzedine
e-mail: Tahar.ezzedine@enit.rnu.tn

© Springer International Publishing AG 2018   175
R. Lee (ed.), *Computer and Information Science*, Studies in Computational
Intelligence 719, DOI 10.1007/978-3-319-60170-0_13

home automation. However, their work lacks the most important part, namely the interception and use of data in real time. The reference [6] is interested in the use of SOA architecture via a REST API on each sensor node in the WSN. Similarly in [7], a study devoted to the implementation of a system based on the SOA architecture with Web services interoperable via the simple object access protocol (SOAP) directly on the nodes and not on the gateways. A major challenge in the deployment of Web service technology on resource-limited sensor nodes leads to unacceptable overheads in terms of RAM, CPU, bandwidth load, and energy [8]. However, a gateway device is required to support the scalability of management operations, as well as reliability and survivability of the WSN with a minimum of complexity and low maintenance cost. In our scheme, we implemented a climate monitoring system based on SOA [9], which allows to intercept sensor data in real time via MQTT and to index them, analyzes them in memory database [10], and transmits in real time a Web application [11] via Web Socket. The spatial and temporal conceptual modeling methodology in GeoUML [12] makes it possible to design a database for structuring and archiving spatial, temporal, and spatiotemporal data coming from the sensor network. The archived data are exposed via a Web services API (REST, GWS) allowing exploitation and analysis according to several spatial and temporal criteria in different graphic and table formats. Thus, the exploitation of spatial and temporal data makes it possible to generate various layers enabling to represent a phenomenon variation over time in various geographic areas (e.g., variation of the temperature during one year in the Saharan zones). This paper is structured as follows: Sect. 2 is devoted to the presentation of climate monitoring systems, as well as the techniques for data collection, data storage, conceptual modeling of spatial temporal data, and query processing. Section 3 discusses the architecture implemented in our system. The client interfaces of our application are represented in Sect. 4. Finally, Sect. 5 concludes the paper.

2   Climate Monitoring System

The climate network sensors based on monitoring systems are mainly composed of a set of sensor nodes (agile or fixed or mobile). This type of network has usually many sensor nodes distributed over an area defined to measure a physical quantity or track events that communicate with each other. Each node in the network is considered intelligent. It is equipped with an acquisition module which provides a measurement of climate data (such as temperature, humidity, pressure, and sunshine.), a processing capacity, a storage volume, a communication system, and energy. Sensors can be used and placed wherever one needs to collect useful information. These systems are responsible of technical collection and information saving. There are two types of sensor data:

- Static data that are related to sensor characteristics (such as its type, storage capacity, and its transmission power.).

- Dynamic data that are collected in relation to the physical features of the environment (such as temperature, humidity, and light.). Their values depend on the variation of the observed phenomenon [13]. Thus, the data collection position varies in case of agile or mobile sensors.

Collected data from the sensors can be spatial, temporal, or spatiotemporal. Indeed a spatiotemporal conceptual modeling is necessary to structure the various data in a central database.

2.1 Collecting Data

Data collection is a basic operation in the WSN. It consists of transmitting the data collected by sensors to a base station. Generally, data collection can be triggered by requests (at the request of the user) for information on the network. It can also be periodically triggered to monitor a geographic area, as it can be triggered by the occurrence of an event. The data collected at the base station can be processed immediately to meet the needs of the user. They can also be stored in a database for analysis by the following [14]. There are two types of data collection: (i) the data collected and transmitted individually to the base station when the measures are important and (ii) the data aggregation, defined as the data fusion process from several sensors at intermediate nodes in order to eliminate redundant transmissions and to reduce the amount of data transmitted across the network [15]. The data aggregation seeks to collect the most critical sensors and transmit them to the base station in effective way with a minimal latency and a low power consumption [16].

2.2 Data Storage

There are also two data storage approaches in WSN [17]. (i) Storage approach which is based on a centralized system. The data are collected from a sensor array and sent to a centralized database. Data processing is performed at the base to respond to user requests. In this approach, the processing of applications and the access to the sensors network are separated [18]. The storage approach is well suited to meet predefined queries on historical data. (ii) Distributed approach, where the data is stored in a database server and the sensors themselves, which will be then considered as databases. Generally, the database system builds an implementation plan for the treatment of requests [19]. The sensors storage capacity is limited and as the database system they form must maintain a history, it is necessary to clean the sensors database simultaneously with measurement acquisition. Indeed, the storage of sensors data in the central database is the adopted solution to climate monitoring system.

2.3   Spatiotemporal Conceptual Model

Spatiotemporal conceptual models aim to answer the users who need to answer their needs and to enable them to describe a diagram of readable and easy data to apprehend. The installation of an information system requires data modeling. If it is not contestable to make a traditional modeling of data in unified modeling language (UML) [20], it is more difficult to find a standard modeling for spatial data, and even for spatiotemporal data that are geometrically different. The spatiotemporal conceptual model must capture the essential semantics of the information change over time. It should be compatible with the traditional model in order to allow the modeling of data that are neither spatial nor temporal [21]. In a spatiotemporal conceptual model, we should know the various types of links (conventional, spatial, and temporal) between entities in the real world. However, there are various spatial conceptual models of data (CMD) among which we state two different approaches: Perceptory [22] and modeling of application data with spatiotemporal (MADS) [23]. They are based on CMD that originates from the databases, respectively, UML and ER (Entity/Association). They are extended to the spatial concepts. Considering the limitations related to MADS [24], we chose to use the Perceptory tool. This tool, which is able to be integrated into the Engineering Software Workshop (ESW) Visio, offers effectiveness in the visual modeling of spatial databases (DBS) and the basic of spatiotemporal data [22]. Perceptory was developed starting from a standard directed-object formalism. Later, it was undergone and was enriched to support the spatial reference and to take into account the norms ISO-TC211. In this paper, we applied spatial modeling GeoUML with the Perceptory tool using as a model real-time spatiotemporal data model [12]. The model mainly describes the aspects of environment monitoring. In this real-time spatiotemporal data (RTSTD) model, we define a GeoUML representation of the spatiotemporal data in a real time and their relationship with other static entities. The data collected by a sensor node are called according to their locations and their dates of acquisition. The sensors are located in the same area of observation to allow collection of various types of data (temperature, pressure, humidity, etc.) [25].

2.4   Queries Processing in WSN

The data acquired by the sensors are localized and dated. They are used in real time and deferred time managing a variety of large masses of spatiotemporal data. Several types of queries can be applied to these data, to spatial, temporal, and spatiotemporal queries. We can mention examples of requests processed by our system:

- Spatial Queries: they seek the value of a product attribute in one location. For example, the query "SELECT temperature FROM sensors WHERE sensors-location = (50, 60)" seeks the temperature of a sensors located in a point in the coordinate plane (50, 60).

- Time Queries: they seek the value of a product attribute at a certain date. For example, the query "SELECT temperature FROM sensors WHERE mesure_date = '10/10/2016'" searches the temperature of each sensor on the mesure date of '10/10/2016'.
- Spatiotemporal Queries: they seek the value of a product attribute at a time and in a given location. For example, the query "SELECT temperature FROM sensors WHERE mesure_date = '10/09/2016' and sensors_location = $(50, 60)$". Researches localized temperature sensor (50, 60) at the mesure date of '10/9/2016'.

3 System Architecture

In this work, we adopted a solution based on a SOA architecture interconnected with MQTT which allows the integration of the real physical peripherals data in other information systems. This architecture is based on the use of Web services technologies (REST, GWS) and the Web sockets with the components of customer RIA. The emergence and the consolidation of service-oriented architecture (SOA), MQTT, and of the wireless sensor network (WSN) increase the advantages (such as flexibility, evolutionarily, the security, interworking, and the adaptability) of this kind of applications. The following architecture in Fig. 5 summarizes the relations between all technologies we have cited. This architecture is made up of the three following modules.

3.1 Module 1

A wireless sensor network (WSN) composed of a large number of nodes where each is equipped with an Xbee module and a set of sensors to detect various physical phenomena such as the sunning, the temperature, and the pressure. The data collected by the various sensors are sent toward a gateway via Wifi connection. The Wifi connection is configured between the nodes and the gateway by using a Zigbee protocol. The communication of information between the gateway and the application server is based on the use of MQTT protocol which consists of a communication channel publish/subscribe Fig. 1.

3.1.1 Real-time Data Exchange

The framework is based on a service-oriented architecture (SOA) and a CloudMQTT allowing a transparent integration of various data collected toward a server (broker). The MQTT is responsible for making the connection between the gateway and the application server or different customers registered in the topic. The models of

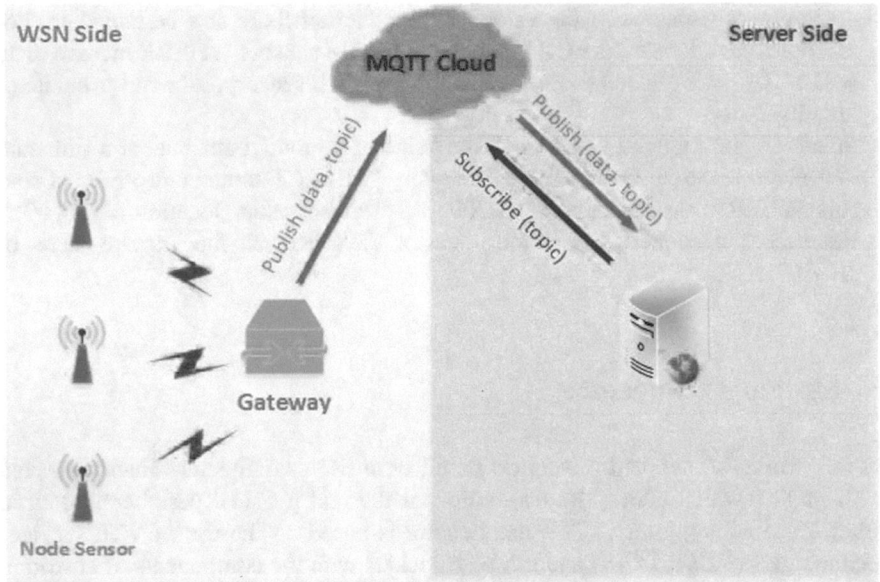

Fig. 1 Mechanism of transferring data between the gateway and the application server

communication publish/subscribe provide a form of interaction which performs less dependency between the entities on the systems. The basic idea is that the application server fits in an object (topic) or a set of events, when the gateway generates events or publishes data on the broker [26], and the latters will be propagated in real time in the server.

3.1.2 REST Architectural Style

The configuration of the sensor network through the clients is done by consuming a service between the gateway and the application server Fig. 2. It allows client applications to reconfigure and maintain the network during its lifetime. So we defined a representational state transfer (REST) style SOA to allow integration of the WSN with the application services REST-WS [27]. REST is an architectural model that triggers a good extension of Web services by presenting various Hypertext Transfer Protocol (HTTP) methods and operations to transfer representations of resources between clients and servers. REST is less complex than SOAP-WS [28], and its consumption is minimal, which provides an open and flexible framework, allowing evolutionary and dynamic monitoring of resources [29, 30]. The format used for data transfer in our work is JSON. Unlike XML, JSON provides a more readable and easy-to-implement structure involving less use of application and network resources.

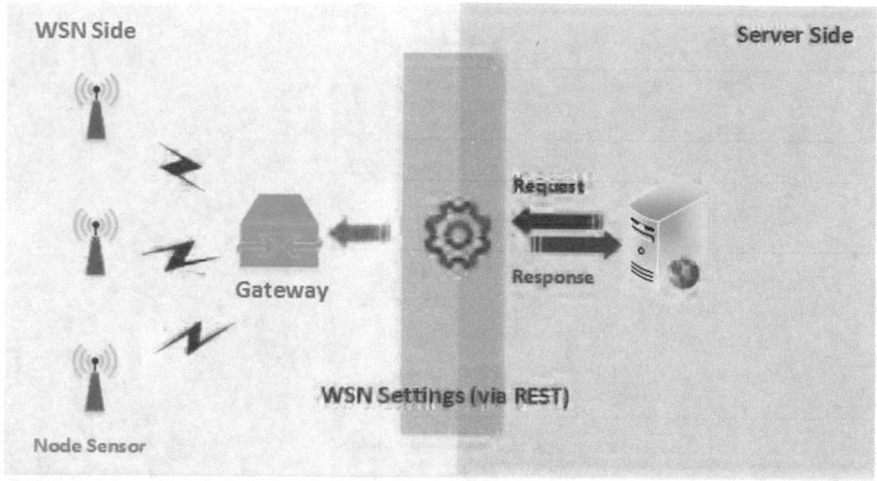

Fig. 2 Mechanism of the sensor network configuration

3.2 Module 2

The application server (which complies with the industry standard) enables the reception of data sent by the gateway and the distribution of users and information loads while ensuring high availability. The module is also based on the use of a database in "In Memory Database". This allows data to be saved and indexed in real time to optimize the backup time and speed up access to recent data. Then, this data will be archived in a PostgreSQL/PostGIS geographic database server which allows the storage, archiving, and management of spatiotemporal data according to well-defined periods. A cartographic server certified by the Open Geospatial Consortium (OGC) complying with ISO/TC 211 standards [31] is thus used for manipulating the display and distribution of geographic data in a Web environment via GWS Web services, WMS). The data received by the application server is transmitted to the clients in real time via Web sockets Fig. 3. The notifications are sent in "Push" mode to the client without execution of the requests when the server receives the new data.

The manipulation and consultation of the data stored in memory database and the spatiotemporal database are done by consuming a service between the clients and the application server via HTTP Fig. 4.

3.3 Module 3

The Web client module allows secure access to the different data. It enables thin clients to query the spatiotemporal or the memory database and then to view the network data of the sensors in a Web environment. This module is based on the

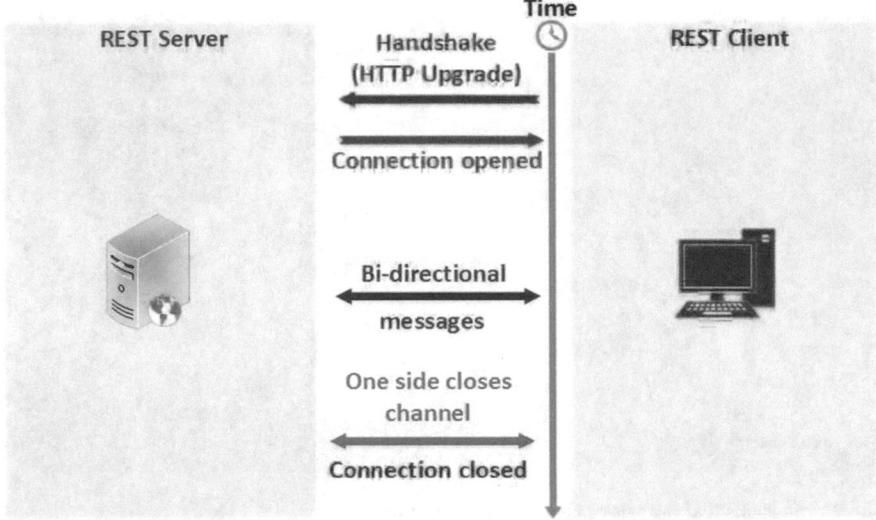

Fig. 3 Communication between client and application server

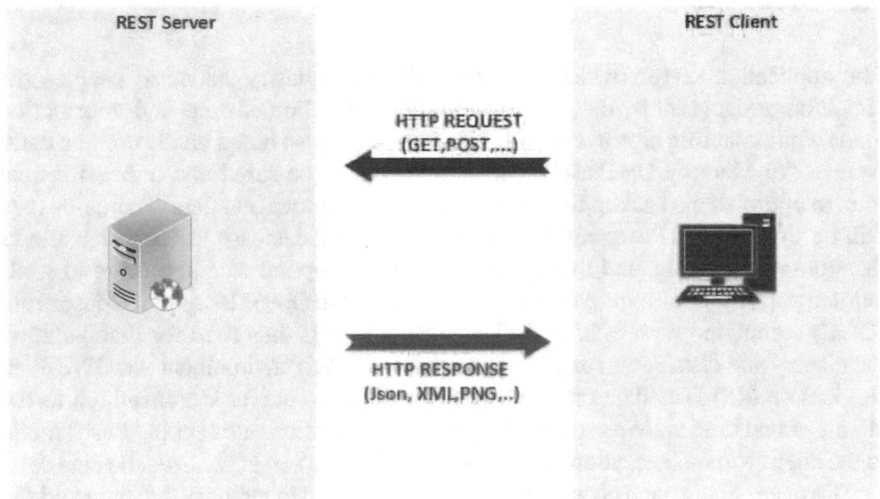

Fig. 4 Query operations to application server

use of RIA Web client available through Web browsers and mobile devices such as smartphones and desktops. Thus, the data stored in the spatiotemporal database can be interrogated and manipulated by heavy clients, such as open source software (Qgis, Gvsig, Quantum GIS) as well as Arcgis, which allows to organize, manage, analyze, communicate, and disseminate geographic information.

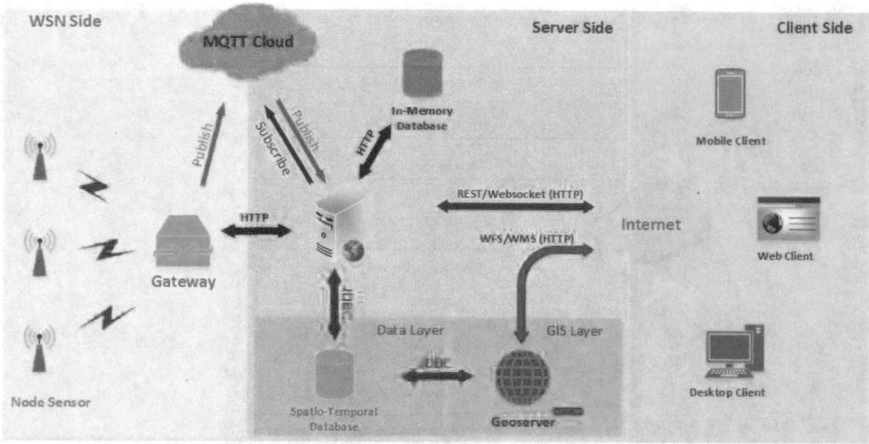

Fig. 5 Architecture patterns of the system

4 System Implementation

The implemented system can intercept the data of WSN. The received data are immediately processed by a filter and treatment modules. Thus, interesting data (critical value or abnormal phenomenon) triggers an alert or a notification that will be sent to users in real time. Filtering and processing modules allow the backup of data in the memory database to ensure high availability system. A dashboard was made available to the user for monitoring various collected data and phenomena Fig. 6. Other interfaces detail the observations of each phenomenon.

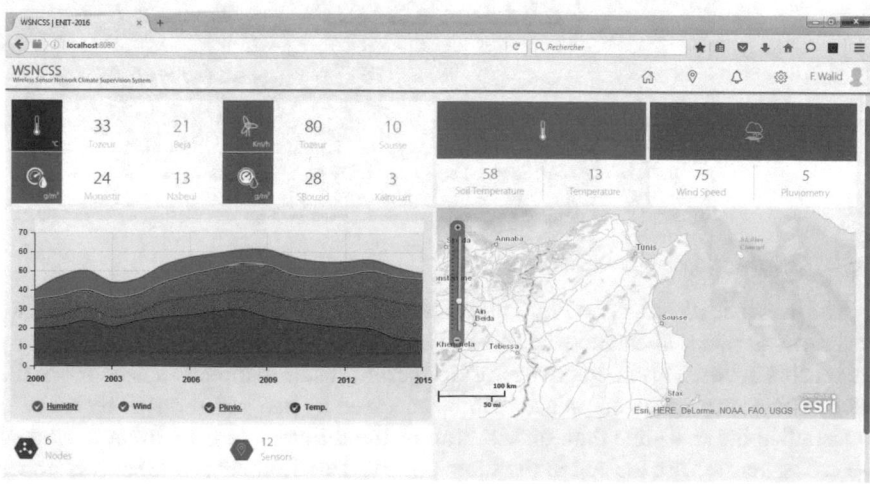

Fig. 6 Proposed system dashboard

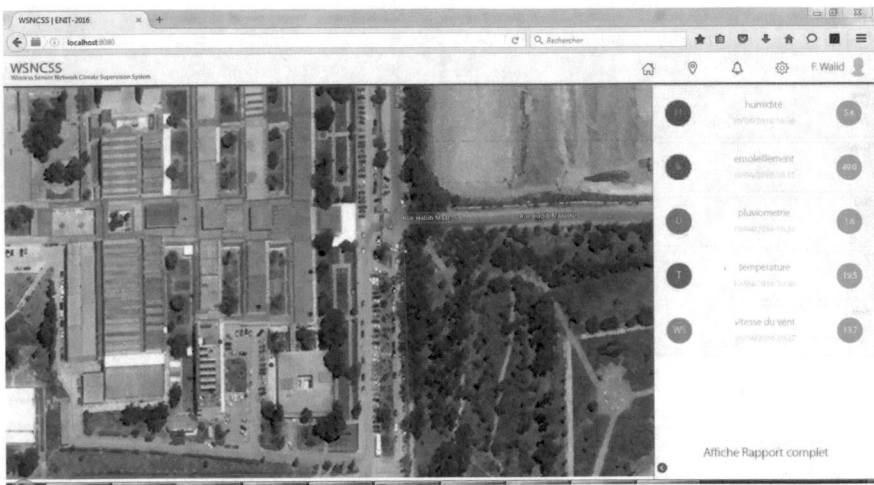

Fig. 7   Interface consulting real-time information sent by the sensor

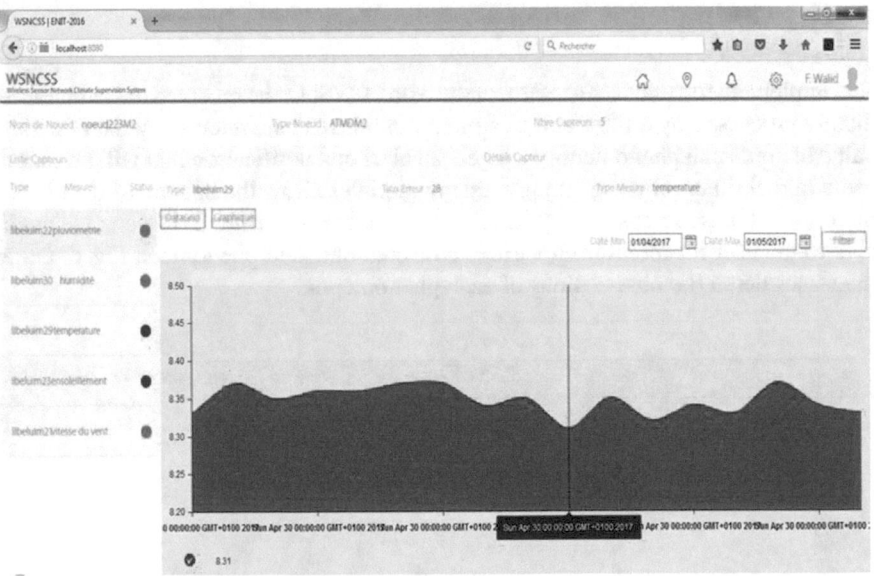

Fig. 8   Chart of temperature measurement

The interface illustrated in Fig. 7 allows visualizing in a Web environment the data of sensors network. It enables the users to consult the different observations sent by the sensors nodes. This interactive interface permits users to directly select one of the sensors on a map. One node sensor can be accessed by pan and zoom or by browser to visualize the real-time data distribution of the different sensors in an attributed table. The sensors are located in the same observed environment. The sensors nodes offer the possibility to measure different variables such as air temperature, rainfall, humidity, and solar radiation. Indeed, several measurements can be performed by a single sensor node.

The interactive interface presented in Fig. 8 allows users to view the collected data in the form of a graphical representation. By clicking on the graphic button, graphics display the data variation in a selected period. The data can be displayed in a data-grid representation mode if the tab button has been clicked.

5 Conclusion

Our application is intended to provide a tool enabling risk specialists to carry out expertise on events, sometimes on recent data in real time or very old data which affected the territory by confronting various sources. These expertises, called summary, will be used by risk managers to refine and improve their decision making, referring not only on the size of the harvested measures but also on the geographic location of events. Finally, the sensor data are used to supply a decision support system. Our application is a platform based on Web services (WFS, WMS, and REST) dedicated to the exchange and sharing of data and cartographic information in the field of natural risks.

References

1. Percivall, G., Reed, C.: Sensor web enablement standards. Sensors Transducers J. **71**(9), 698–706 (2006)
2. Hefeeda, M., Bagheri, M.: Forest fire modeling and early detection using wireless sensor networks. Ad-Hoc Sensor Wirel. Netw. **7**(3–4), 169–224 (2009)
3. Cortes, C., Fisher, K., Pregibon, D., Rogers, A.: Hancock: a language for extracting signatures from data streams. In: Proceedings of the Sixth ACM SIGKDD International Conference on Knowledge Discovery and Data Mining, p. 917. ACM Press, New York (2000)
4. Cranor, C., Gao, Y., Johnson, T., Shkapenyuk, V., Spatscheck, O.: Gigascope: high performance network monitoring with an SQL interface. In: Proceedings of the ACM SIGMOD International Conference on Management of Data, p. 623. ACM Press, New York (2002)
5. Malatras, A., Asgari, A., Bauge, T.: Web enabled wireless sensor networks for facilities management. IEEE Syst. J. **2**, 500–512 (2008)
6. Baccar, S., Rouached, M.: A web services based approach for resource-constrained wireless sensor networks. IJCSI Int. J. Comput. Sci. Issues **9**(3, No 2) (2012)
7. Kyusakov, R., Eliasson, J., Delsing, J., van Deventer, J., Gustafsson, J.: Integration of wireless sensor and actuator nodes with IT infrastructure using service-oriented architecture. IEEE Trans. Ind. Inf. **6**(1) (2011)
8. Priyantha, N.B., Kansal, A., Goraczko, M., Zhao, F.: Tiny web services: design and implementation of interoperable and evolvable sensor networks. In: SenSys 08: Proceedings of the 6th ACM conference on Embedded Network Sensor Systems, New York, NY, USA, pp. 253–266, 4–7 Nov 2008
9. Fantazi, W., Ezzedine, T., Bargaoui, Z.: Implementing a sensor network for monitoring of drought indicators. In: International Scientific Symposium of Water Management and Desertification, Istanbul/Turkey, 26–28 Nov 2014

10. Fantazi, W., Atoui, A., Ezzedine, T.: Captree: spatial and temporal indexing in databases from fixed sensors. In: Proceedings of 2016, 4th International Conference on Control Engineering and Information Technology (CEIT-2016), Tunisia/Hammamet, 16–18 Dec 2016
11. Chris, L.: Rich Internet Applications: Design, Measurement, and Management Challenges, Keynote Systems (2006)
12. Fantazi, W., Ezzedin, T., Bargaoui, Z.: Wireless sensor network to help low incomes to face drought impacts. Int. J. Comput. Electr. Autom. Control Inf. Eng. **9**(8) (2015)
13. Zhao, F., Guibas, L.J.: Wireless Sensor Networks: An Information Processing Approach. Morgan Kaufmann Publishers (2004)
14. Incel, O., Ghosh, A., Krishnamachari, B.: Scheduling algorithms for tree-based data collection in wireless sensor networks. In: Theoretical Aspects of Distributed Computing in Sensor Networks, pp. 407–445. Springer, Berlin/Heidelberg (2011)
15. Ozdemir, S., Xiao, Y.: Secure data aggregation in wireless sensor networks: a comprehensive overview. Comput. Netw. **53**(12), 2022–2037 (2009). Aug
16. Li, H., Wu, C., Yu, D., Hua, Q., Lau, F.C.M.: Aggregation latency-energy tradeoff in wireless sensor networks with successive interference cancellation. IEEE Trans. Parallel Distrib. Syst. (TPDS) **24**(11), 2160–2170 (2013)
17. Diallo, O.: Joel J.P.C. Rodrigues and M. Sene.:Real-time data management on wireless sensor networks: A survey. Journal of Network and Computer Applications **35**(3), 1013–1021 (May 2012)
18. Bonnet, P., Gehrke, J., Seshadri, P.: Towards sensor database systems, ACM Digit. Library 3–14 (2001)
19. Li, C., Zhang, H., Hao, B., Li, J.: A survey on routing protocols for large-scale wireless sensor networks. Sensors **11**(4), 3498–3526 (2011)
20. Booch, G., Rumbaugh, J., Jacobson, I.: Le guide de lutilisateur UML, collection Technologies object/Reference, Paris/Eyrolles, p. 534 (2000)
21. Parent, C., Spaccapietra, S., Zimnyi, E., Donini, P., Plazanet, C., Vangenot, C., Rognon, N., Rausaz, P.: MADS: modle conceptuel spatio-temporel. Revue Internationale de Géomatique **7**(3–4), 317–351 (1997)
22. Bédard, Y.: Visual modelling of spatial databases: towards spatial PVL and UML. Géomatica **53**(2), 169–186 (1999)
23. Spaccapietra, S., Parent, C., Zimnyi, E.: Modeling time from a conceptual perspective. In: Proceedings of the 7th International Conference on Information and Knowledge Management, Bethesda, Maryland, USA, pp. 432–440 (1998)
24. Laplanche, F.: Conception de projet SIG avec UML. Bulletin de la Socit gographique de Lige **42**, 19–25 (2002)
25. Fantazi, W., Ezzedine, T., Bargaoui, Z.: Conceptual modelisation of spatio- temporal database based on wireless sensor network to follow drought indicators. In: International Conference on Applied Geology and Environment (ICAGE 2016), Tunisia/Mahdia, 19–21 May 2016
26. Lampkin, V., et al.: Building Smarter Planet Solutions with MQTT and IBM WebSphere MQ Telemetry (Sep 2012). doi:268/s0738437085
27. Richardson, L., Ruby, S.: RESTful Web Services. OReilly (2007)
28. W3C Consortium: W3C SOAP 1.2 specification (2007)
29. Pautasso, C., Zimmermann, O., Leymann, F.: Restful web services vs. "Big" web services: making the right architectural decision. In: Proceedings of ACM 17th International Conference on World Wide Web (WWW), China/Beijing, Apr 2008
30. Landre, W., Wesenberg, H.: REST versus SOAP as architectural style for web services. Paper presented at the ACM SIGPLAN International Conference on Object-Oriented Programming Systems, Languages, and Applications, Montreal/QC/Canada, Oct 2007
31. Yuan, Y., Zheng, W.: The WebGIS development base on GeoServer's. Softw. Guide **1**(3), 96–98 (2007)

Mobile Application Development on Domain Analysis and Reuse-Oriented Software (ROS)

Mechelle Grace Zaragoza and Haeng-Kon Kim

Abstract The term reuse is suggested to be a key to improving of any software development and productivity, particularly where one can identify a family of systems. As to mobile application development, one should consider not only component development, but also a classification for reusable domain pertaining to software. We classify the domain component considering functional and non-functional factor identified through domain analysis. This paper briefly describes the study of mobile application development with the use of domain analysis. How to reuse software will cause certain advantages and issues upon reviewing this study. Also, we proposed a flowchart based on the reuse method of large-scale embedded software based on inter-module relations on process flow of the proposed reuse method.

Keywords Mobile application · Domain analysis · Reuse-oriented software

1 Introduction

The number of demands the mobile application industry gets every single day is rapidly demanding. As it continues to burst, developers need a great part of their skills to develop mobile application in a limited time, but maintain good and high-quality mobile application products to satisfy the needs of their customers.

Nowadays, as demand for software is growing, the lines of code that must be developed tend to increase; moreover, much shorter development time is required [1]. Recorded currently, application development for smart devices is an evolving field with great economic and scientific interest. [Gartner] as the total number of mobile app used and is being downloaded worldwide will eventually increase to 81

M.G. Zaragoza · H.-K. Kim (✉)
Catholic University of Daegu, Gyeongsan, South Korea
e-mail: hangkon@cu.ac.kr

M.G. Zaragoza
e-mail: mechellezaragoza@gmail.com

© Springer International Publishing AG 2018
R. Lee (ed.), *Computer and Information Science*, Studies in Computational
Intelligence 719, DOI 10.1007/978-3-319-60170-0_14

billion in 2013, and paid downloads will surpass 8 billion and free downloads 73 billion [2]. Collected facts about mobile development market is dominated by these five big platform providers, namely as to Nokia with its Symbian OS in (46.6%), 2 Apple with its iPhone OS in (17.3%), RIM with its Blackberry OS (15.2%), Microsoft with its Windows CE OS family in (13.6%), and LiMo Foundation with its Linux Mobile operating system in (5.1%). Google recently launched its Android operating system and more likely to become part of the big players in the industry [3].

1.1 Domain Analysis

By all means, domain analysis is a process by which information is used in developing software systems that identifies, captures, and organizes with the purpose of making it reusable when creating systems that are new. During software development, several kinds of information are generated, beginning with requirements analysis to specific designs to source code.

The said source code is at the lowest level of abstraction and considers the most detailed representation of a software system. Complementary key information is also generated during software development. This code documentation, history of design decisions, testing plans, and user manual are essential to take a better understanding of the entire system. However, one must consider the advantages and the disadvantages of using domain analysis in recycling old mobile applications that are very rampant as the emergence of mobile applications is at its peak of popularity [4].

1.2 Reuse-Oriented Software Engineering

The term reuse is one of the major concepts of today's software engineering since this can not only save an amount of work when existing components providing a given functionality are reused, but this can also help existing components that might have lots of testing received so far so we can possibly build more reliable systems based on them. Today, the number of software projects applies reuse to some extent; however, some of them relies more on reused components than others.

And oftentimes, reuse happens in an informal way, as it simply means copying the previous project, code, designs, and requirements. They try to see what is beyond these, modify. and incorporate into the system. This is basically the application of patterns in the development process.

Figure 1 shows a general process model for reuse-oriented software engineering. Initial requirements specification stage and the validation stage are compared with other software processes; the intermediate stages in a reuse-oriented process are different.

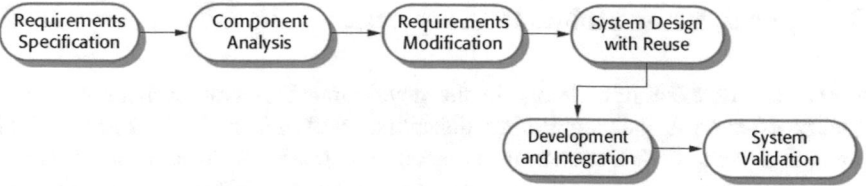

Fig. 1 Process model for reuse-oriented software

[SOMMERVILLE 2010], these include 4 stages:

1. Component analysis. Basing on the requirements specification, components that implement the specification. Mostly, there is no exact match and the components may be used only to provide some of the functionality that is required.
2. Requirements modification. A stage where the requirements are analyzed using information about the components that have yet been discovered. They are then modified to and reflected to the available components. Where modifications are impossible, the component analysis activity may be reentered to search for other solutions.
3. System design with reuse. A stage where the framework of the system is designed or an existing framework is reused. Developers will perform the design by taking into account the components that are reused, and they will organize this framework accordingly. A new set of software may have to be designed if reusable components are unavailable.
4. Development and integration. Final stage where software that cannot be externally obtained is developed, and the components and commercial off-the-shelf (COTS) systems are integrated to create a brand new system. System integration, in this model, could be a part of the development process rather than a separate activity [5].

2   Related Work

This work is related to the improvement of the process in the software reuse development of large-scale embedded software which reuses modules from multiple systems. In this study, holding development, management, property information individually corresponding to source files, modules and software blocks and using the information in the reuse development flow, management, and development process are improved. By this method, a source file is analyzed first, then the dependency on the symbol level between modules is extracted and the functional hierarchical structure of the software is made visible in the form of a block diagram [1].

3 Challenges in Mobile Development

Recent use of existing solutions in the development of new systems is a main quality of every good engineering discipline. Software reuse is a state of the practice development approach in application domains, such as telecommunications, factory automation, automotive, and avionics. Software engineering has produced several techniques and approaches for promoting the reuse of software in the development of complex software systems [6].

There are a lot of challenges in developing mobile services and applications. There are great variety of mobile standards, operating systems on different devices as one application may work on one cell phone, while it does not on the other [7].

Systematic reuse requires vivid information of previous works, in particular a major problem about the creation of assets that can be reused in a context different from that where they have been. In this view, domain analysis is a fundamental activity integrated in a software process based on reuse.

Numbers of approaches already exist pertaining to domain analysis, but they are not basically that popular. Valid reasons are that they are too challenging and rigid to make and could be that they target assets that do not have a high reuse potential. Mostly, none of them is specifically targeted to design frameworks.

3.1 Domain Characterization and Project Planning

Steps in any domain method are mentioned as a preparatory activity that targets to collect the data the least information about the problem.

Activities mentioned during the phase are the lists:

a. Business analysis
b. Feasibility analysis,
c. Domain description,
d. Project planning and resource allocation.

3.2 Data Analysis

Essential information to the analysis is collected and organized, and then the analysis exploits domain commonalties and variations. Activities mentioned are as follows:

a. Data organization,
b. Data exploitation.

3.3 Domain Modeling

The purpose of the modeling phase is to complete the previous analysis step building suitable domain models of the domain. Here are the lists:

a. Modeling commonalty's aspects in the domain,
b. Refining domain models encapsulating variation possibilities,
c. Defining frameworks and general architecture for the domain,
d. Describing the rationale beneath domain models and tracing technical issue and relative decisions made in the analysis and modeling process.

It is the core activity aiming to produce reusable assets, such as components, framework, and architectures. In comparison, domain modeling is a class of similar systems in a specific application domain, while system modeling is the specific software system that has to be built.

3.4 Evaluation

The role of evaluation is to verify the results of each step of the domain analysis process, classifying possible faults done in constructing the model, and to validate the results against requirements and user expectations [8].

4 Systematic Reuse Needs a Systematic Approach

In order to create good software, one must take note of an institutional organizational approach to produce a product development in the purpose of reusable assets to be created or acquired. By systematic reuse, that means institutionalized organizational approach to product development in which reusable assets are purposely created or acquired, and then consistently used and maintained to obtain valuable levels of reuse, optimizing the organization's ability to create quality software products which are firm and effective [9].

Identifying and achieving software reuse for mobile applications are fundamental problems in software engineering. The current approaches for component reuse concentrate primarily on the consumer perspective for it [10].

5 Proposed Flowchart

Figure 2 depicts the process flow of the proposed method in which new software Product-B is developed by reusing the existing application software: Product-A

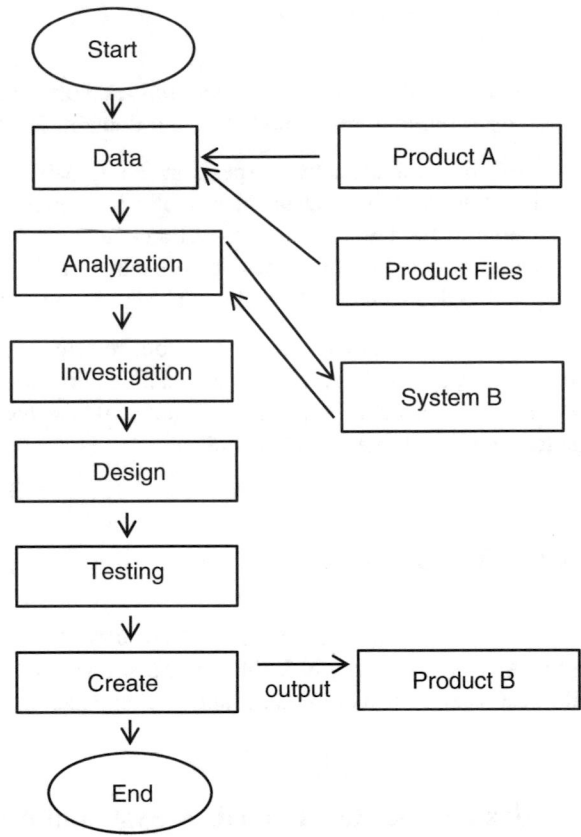

Fig. 2 Reuse process flow by inter-module relation analysis and visualizing associated information

(1) Gathering of data
    Setting up the specifications of Product-B to be developed.
(2) Relation analyzation
    The associated information about the symbols among the files is analyzed.
    Furthermore, the software, structural data, and the software block diagram are
    created for the group of software.
(3) Investigation
    The developer then investigates the reused software modules by visualizing
    their software structure displayed in a block diagram based on relation analysis
    information. The data of all software modules associated with the modified
    module to reuse is displayed in the block diagram.
(4) Design
    In the design phase, the developer performs modification of the existing
    software modules and addition of new software modules.

(5) Testing
   Testing takes place after enormous data gathering and investigation to make sure its usability.
(6) Build
   The developer builds the software through the compile and link steps to create executable code of software Product-B [1].

6 Why Software Reuse Has Failed Historically?

Reuse has been a popular topic of debate and discussion for the past 30 years in the software community. Developers have successfully applied reuse opportunistically. Opportunistic reuse works fine in a limited way for individual programmers or small groups but the fact that it does not scale up across business units or enterprises to provide systematic software reuse. Systematic software reuse is a promising means to eliminate development cycle time and cost and improve software quality, and effort by constructing and applying multiuse assets such as architectures, patterns, components, and frameworks [11].

7 Conclusion

Change is unavoidable even the system that we put into the operating environment of our techy world. Developers only focused on the new things they might come up but lately developers started to recognize the use of achieving better results in using reuse software based on a well-planed systematic approach. This has been tested and proven to reduce the risks of time consumption, cost, and effort. When building a new software, one must first consider the previous system used if it has something to do with the latest one. By doing this, it will save the cost of developing, testing, and documenting and maintaining list of developing software. In order to come up with less programming and efforts as reusing software might be complex, it is very safe to deliberate first the functionality and uses. There have certainly successful reuse software in the market with sophisticated frameworks and components.

Acknowledgements This Research was supported by the MSIP (Ministry of Science, ICT and Future Planning), Korea, under the C-ITRC (Convergence Information Technology Research Center) support program (IITP-2016-H8601-15-1007) supervised by the IITP (Institute for Information & Communication Technology Promotion).

This Research was supported by the International Research & Development Program of the National Research Foundation of Korea (NRF) funded by the Ministry of Science, ICT & Future Planning (Grant Number: K 2014075112).

This Work was partially supported by Catholic University Research Fund in 2017.

References

1. Hidetoshi, K.: A reuse method of large-scale embedded software based on inter-module relations. Softw. Eng. **4**(1), 1–9 (2014)
2. Xanthopoulos, S., Xinogalos, S.: A Comparative Analysis of Cross-platform Development Approaches for Mobile Applications, Conference Paper September (2013). doi:10.1145/2490257.2490292
3. Holzer, A., Ondrus, J.: Trends in mobile application development. In: International Conference on Mobile Wireless Middleware, Operating Systems, and Applications. Springer Berlin (2009)
4. Prieto-Diaz, R.: Domain analysis: an introduction. ACM Sigsoft Software Engineering Notes, vol. 15(2), p. 47
5. Sterbinszky, N.: Case study in system development—notes
6. Brugali, D.: A reuse-oriented development process for component-based robotic systems. In: International Conference on Simulation, Modeling, and Programming for Autonomous Robots. Springer, Berlin (2012)
7. Amatya, S., Kurti, A.: Cross-Platform Mobile Development: An Alternative to Native Mobile Development, 013-10-29
8. Valerio, A., Succi, G., Fenaroli, M.: Domain Analysis and Framework-Based Software Development. Software Production Engineering Lab (LIPS)
9. Griss, M.L.: Systematic Software Reuse: Architecture, Process and Organization are Crucial, Laboratory Scientist. Software Technology Laboratory, HP Laboratories
10. Kim, H.K.: Mobile Application Development Using Component Features and Inheritance, Computer Applications for Software Engineering, Disaster Recovery, and Business Continuity, pp. 53–62
11. Schmidt, D.C.: Why Software Reuse has Failed and How to Make It Work for You

A Transducing System Between Hichart and XC on a Visual Software Development Environment

Takaaki Goto, Ryo Nakahata, Tadaaki Kirishima, Takeo Yaku
and Kensei Tsuchida

Abstract In recent years, embedded systems have been widely used in various fields. However, the development burden of embedded systems tends to be high because they are complex. A visual development environment is one way to reduce this burden. We have already proposed a visual programming development environment for program diagrams called Hichart. In this paper, we describe a visual development environment for the XC language, which is a programming language for XMOS evaluation boards.

1 Introduction

Embedded software is widely used in household electric appliances, cars, bank ATMs, ticket vending machines, and automatic ticket gates at stations. Currently, there is an increasing trend toward increasing the development scale of embedded software while decreasing the development period, owing to networking and enhancement of the functions of embedded devices, diversification of end-user needs, and shortening of product life cycles. If a program becomes large scale and complicated, its contents and flow may become difficult to understand. One solution to this problem is to use program diagrams. The program diagram language Hichart directly reflects the hierarchical tree structure of a program. Therefore, it could be used to immediately understand the overall structure of a program and to greatly reduce the labor involved in creating and maintaining a program.

T. Goto (✉)
Ryutsu Keizai University, 120 Ryugasaki, Ibaraki 301-8555, Japan
e-mail: tg@gotolab.net

R. Nakahata · T. Kirishima · K. Tsuchida
Toyo University, 2100 Kujirai, Saitama, Kawagoe-shi 350-8585, Japan
e-mail: kensei@toyo.jp

T. Yaku
Nihon University, 3-25-40 Sakurajosui, Tokyo, Setagaya-Ku 156-8550, Japan
e-mail: yaku.takeo@nihon-u.ac.jp

© Springer International Publishing AG 2018
R. Lee (ed.), *Computer and Information Science*, Studies in Computational
Intelligence 719, DOI 10.1007/978-3-319-60170-0_15

In this paper, we aim to develop an embedded software development environment by using the program diagram language Hichart. Specifically, we aim to develop an embedded software development environment for XC language [1], which is a programming language for XMOS's [2] evaluation board. Specifically, we aim to realize a conversion function from XC to Hichart and from Hichart to XC.

2 Background

2.1 XMOS

XMOS is an event-driven multicore processor provided by XMOS and a development board equipped with this processor. This processor can realize real-time processing without an OS with software alone. XCore comprises a 32-bit RISC processor, and its hardware specifications are as follows:

1. 8 threads/1 core,
2. 32 channel end/1 core,
3. SRAM capacity limit,
4. Number of input/output ports (depends on the processor series),
5. XLink, and
6. 10 ns timer.

As mentioned in the hardware restrictions (1), each core has a group of 8 registers called threads, and each thread is like a CPU. These threads (T0–T7) are executed using time division.

Because processing is performed using time division, eight programs can work simultaneously without interference. Of course, because of the use of time division, the speed of the RISC core is obtained as 1/number of executed threads × number of clocks of processor. There is no overhead such as task switching in a real-time OS, and a system or a program can be designed without considering having to wait for a task.

As mentioned in the hardware restrictions (5), the XMOS processor has a function called XLink. An XLink connection enables communications between cores. An XLink connection can be used to connect not only to cores inside the processor but also to processors of the same series. However, an XLink connection cannot be used to connect to processors of different series.

The XCore's clock cycle has a timer of 10 ns (6). Time recording and delay control can be realized by using a timer, and periodic pulses can be generated so that it can also be used for controlling external devices.

An XMOS processor, push buttons, and LEDs are mounted on the XMOS evaluation board, and the boards have the minimum peripheral components necessary for development using XMOS.

2.2   XC

The XC language is available as a development language for XMOS processors. It can be used to develop a system with concurrency and real-time properties without using an OS. The XC language was developed based on the C language. The following functions are added to the C programming notation that is frequently used in embedded systems.

1. Parallel control of processes,
2. Time recording and delay control using a timer,
3. Communication by channels between processes,
4. Event function, and
5. Control of external device with input/output port.

Some functions are excluded from the XC language. Unlike in the C language, the XC language cannot perform operations with pointers, floating point arithmetic, and so on. However, by combining both C language development and XC language development, these drawbacks can be addressed.

2.3   Hichart

Our study uses the program diagram Hichart. Hichart is a program diagram methodology that was introduced by Yaku and Futatsugi [3]. It has three key features:

1. A diagram is a tree flowchart that has the flow control lines of a Neumann program flowchart.
2. The nodes of the different functions in a diagram are represented by differently shaped cells.
3. The hierarchy of the data structure represented by a diagram and the control flow are simultaneously displayed on a plane; this distinguishes it from other program diagram methodologies.

Hichart has attracted much research attention. A prototype formulation of attribute graph grammar for Hichart was reported in [4]. This grammar consists of Hichart syntax rules, which use a context-free graph grammar [5], and semantic rules for layout. The authors have been developing a software development environment based on graph theory, which includes graph drawing theory and graph grammars [6, 7]. Thus far, we have developed bidirectional translators that translate Pascal, C, or DXL source into Hichart and vice versa [6, 7]. For instance, the HiChart Graph Grammar (HCGG) was introduced in [8]. HCGG is an attribute graph grammar with an underlying graph grammar based on edNCE graph grammar [9], and it is intended for use with DXL.

2.4   JavaCC and JJTree

JavaCC is a Java-based compiler/compiler developed by Sun Microsystems, USA. It is a program generator that generates Java programs that execute syntax analysis and lexical analysis.

JJTree is a JavaCC preprocessor. JJTree and JavaCC can be used in combination to automatically generate a program that creates a tree structure from a parsing result. By executing this tree structure, it is possible to obtain the result of the intended program. Normally, when relatively complicated grammar is required, compiler knowledge and techniques such as creating intermediate code are essential. However, if we use JJTree and JavaCC, a parsing result is expressed in a tree structure. Therefore, if we can understand the algorithm for processing the data structure of the tree structure, we can create an interpreter with complicated grammar without special knowledge of the compiler design.

For conversion from each source file to Hichart internal data, JavaCC and JJTree are used. When a token is defined and a character string matches a defined pattern, the character string is regarded as a token.

3   Transducing System

In this study, we aim to enable XC language programming using Hichart by adding an XC language compatible function to the existing Hichart development environment. Figure 1 shows an overview of the proposed system. Users visually perform programming by using the Hichart editor.

The Hichart editor has a behavior check window and a behavior table, and it enables development in consideration of the physical parameters for embedded software development. Programs entered in the Hichart editor are converted to internal data. From internal data, they can be converted to various languages such as C, NXC, NQC, and XC, and they can output a source file in each language as a conversion result. It is also possible to input C, NXC, NQC, and XC source files in the Hichart editor. Source files loaded into the Hichart editor can be converted to Hichart's internal data format. From these, it can be seen that mutual conversion between C, NX, NQC, and XC can be performed using the Hichart editor.

3.1   External Specification

The inputs to this system are C, XC, NQC, and NXC source files. Because XC's grammar is mostly the same as that of the C language, we used the basic structure of the existing Hichart editor and added to and improved it in the system. The subsets added in this study are shown in Table 1.

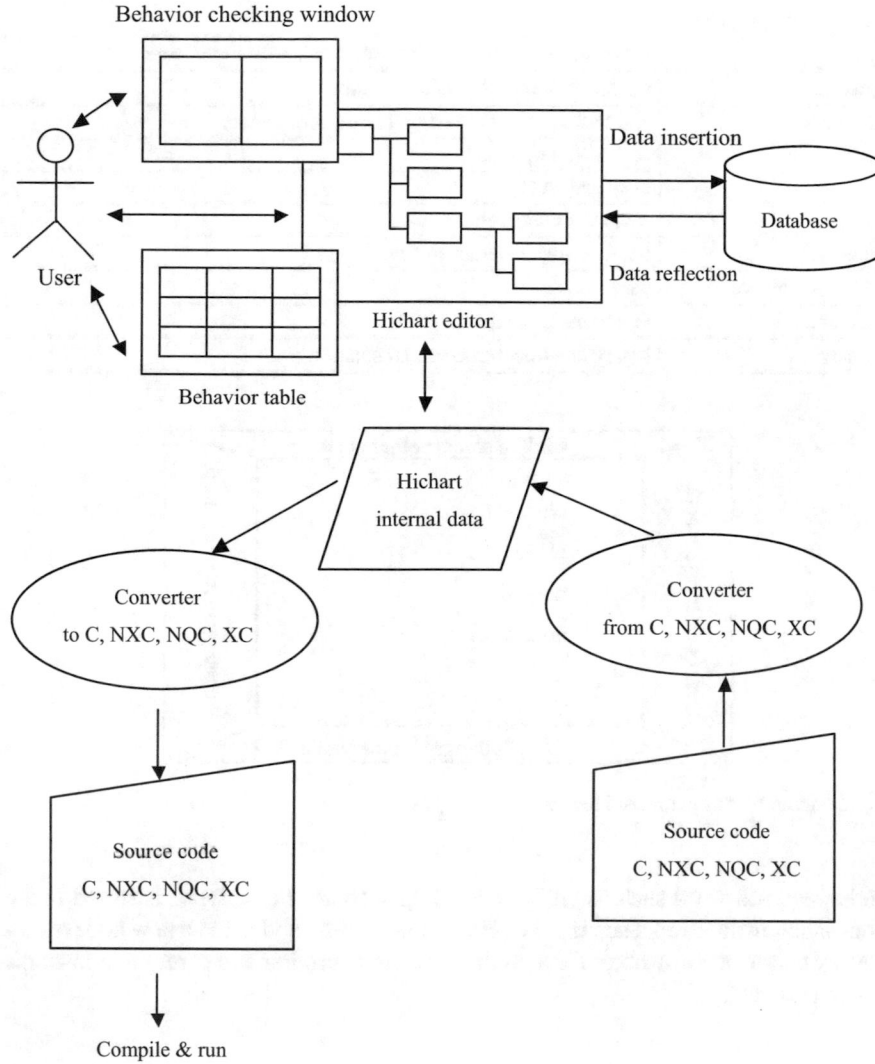

Fig. 1   System overview

3.2   *Internal Specification*

To process Hichart on the computer, Hichart internal data is defined. This definition is performed in the HichartNode class, and its structure is shown in Fig. 2.

Each node has a unique integer node ID. The node type stores an integer defined for each type of node. The node size is defined by the Dimension class, which holds the width and height of a node. An expression or a type name is stored in the string,

Table 1 Added subsets

Basic type	short, signed, unsigned, chan, chanend, port, timer, core
Qualifier	in, out, buffered, streaming, volatile
Operator	:>, :>>>, <:, <:>> (input / output operator)
	&, |, ^, ~(bit operator)
	! (logical NOT)
	<<, >> (shift operator)
Sentence	input sentence, output sentence
Control sentence	par
Declaration	prototype, on sentence
Others	binary notation, hexadecimal notation

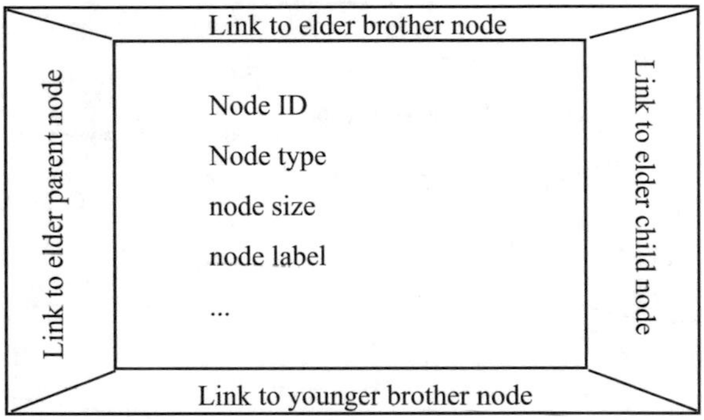

Fig. 2 Inner data structure of Hichart

and a reserved word such as "if" or a heading such as "Func_type" is stored in the node label. In addition, each link is a HichartNode-type node, and a new node is created by calling a constructor for it; at the same time, brother and parent relationships are constructed.

3.3   Conversion Process

For conversion from each source file to Hichart internal data, JavaCC and JJTree are used. When a token is defined and a character string matches a defined pattern, the character string is regarded as a token. Figure 3 shows an example of token definitions.

When the parser detects a character string "if," it is regarded as the token <IF>, and when it detects "+," it is regarded as the token <ADDOP>. Regular expressions can also be used as defined in <IDENT>. We perform parsing using these tokens.

```
|      <IF:        "if">
|      <ADDOP:     "+">
|      <IDENT:     ["a"-"z", "A"-"Z", "_"](["a"-"z", "A"-"Z", "_", "0"-"9"])*>
```

Fig. 3 An example of token definitions

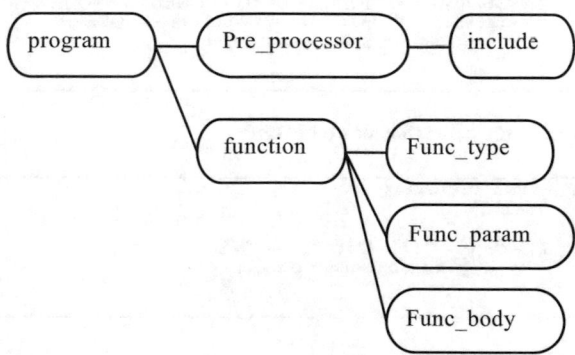

Fig. 4 An example of a syntax tree before conversion

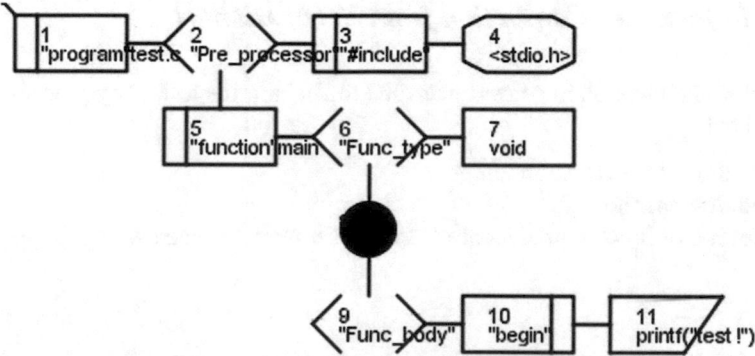

Fig. 5 An example of a syntax tree after conversion

The node descriptor # is used to make tree structure close to the Hichart diagram when generating the syntax tree using JJTree. The implementation of the conversion from the JJTree syntax tree to the HichartNode is performed by traversing syntax trees.

Actions are implemented in SimpleNode.java, which is automatically generated by JJTree. The class of each node (AST ***. Java) inherits it.

Figures 4 and 5 show the syntax tree before and after conversion, respectively.

```
void baseType() : {
        Token t;
} {
        ((t = <STATIC>|t = <CON>|t = <VOLATILE>|t = <IN>|t = <OUT>
        |t = <BUFFERED>|t = <STREAMING>){
                jjtThis.tokenString += t.image + " ";})*
        (t = <INT> | t = <CHAR> | t = <VOID> | t = <DOUBLE> | t = <FLOAT>
        | t=<LONG> | t = <TASK> | t = <SUB> |t = <START> |t = <STOP>
        |t = <SHORT> |t = <SIGNED> |t = <UNSIGNED> |t = <CHAN>
        |t = <CHANEND>|t = <PORT>|t = <TIMER>|t = <CORE>){
                jjtThis.tokenString += t.image;
        }
}
```

Fig. 6 Definition of a production rule for a basic type

```
void unaryOperator() #OPE: {
        Token t;
} {
        ( t = "&" | t = "*" | t = "" | t = "!" | t = "~"){
                jjtThis.tokenString = t.image;
        }
}
```

Fig. 7 Definition of a unary operator

3.4 Function of Conversion from XC to Hichart

To realize the function to convert from XC to Hichart, the following procedure was carried out.

1. Addition of basic type/modifier,
2. Addition operator, and
3. Addition of input, output, control, and declaration statements.

3.5 Function of Conversion from Hichart to XC

The conversion from Hichart to the source file is performed by tracing the node of Hichart and writing out the information for that node. Although there are exceptions to the order of following nodes, basically, it is performed using depth (parent–child relationship) priority (Figs. 6, 7, and 8).

```
void input() : {
        Token t;
} {
        declarator()#when [ t = "@"{jjtThis.tokenString = t.image;} E() ]#AT
        [ t=<WHEN>{jjtThis.tokenString = t.image;} funcCall() ]
        inOpe() ( LOOKAHEAD(2)baseType()declarator()
        | t=<VOID>{jjtThis.tokenString = t.image;} | declarator() )#DVR
        [ t = "@"{jjtThis.tokenString = t.image;}
        ( LOOKAHEAD(2)baseType()declarator()
        | t=<VOID>{jjtThis.tokenString = t.image;} | declarator() )#DVR ]#AT
}
void inOpe() #OPE :{
        Token t;
} {
        t=":>"{jjtThis.tokenString = t.image;}
        [ t=">>"{jjtThis.tokenString += " "+t.image;} ]
}
```

Fig. 8 Definition of production rule for input

Fig. 9 An example: Output
a signal to the port of XK-1
and blinks eight LEDs in
sequence at a "specified time
interval"

```
#include <platform.h>
#define PWM_PERIOD 1000000

on stdcore[0] : out port p_out = XS1_PORT_8A;

void pulse(chanend);
void leddrv(out port,chanend);

int main(void) {
        chan ch;
        par {
                on stdcore[0] : pulse(ch);
                on stdcore[0] : leddrv(p_out,ch);
        }
        return 0;
}

void leddrv(out port p_lrm,chanend ch0)
{
        unsigned led = 0x01<<3 ;
        unsigned signal;
p_lrm <: led;
        while(1)
        {
                ch0  :> signal;
                p_lrm <: led;
                led = (led >> 7 & 0x1) | (led << 1 & 0xFF);
        }
        return;
}

void pulse(chanend ch){
        timer t_mot;
        unsigned time;
        while(1){
                t_mot :> time;
                t_mot when timerafter(time+PWM_PERIOD) :> time;
                ch <: 1;
        }
}
```

Fig. 10 Execution of example (Fig. 9)

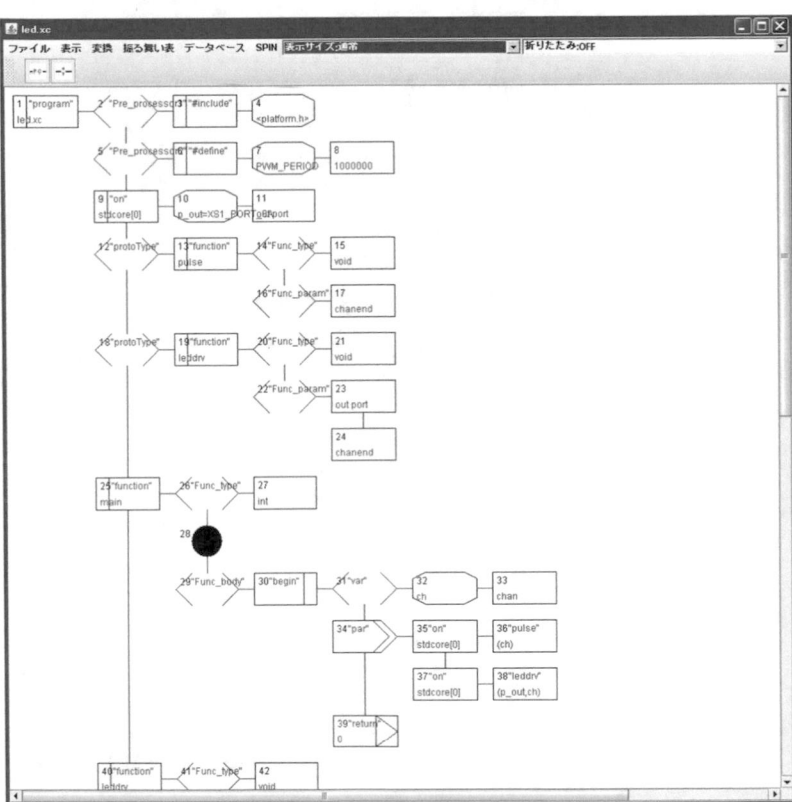

Fig. 11 Result of sample program (1): Hichart editor screen

Fig. 12 Result of sample program (2): XC source file from Hichart

```
XC Editor
File  Edit  Hichart
#include<platform.h>
#define PWM_PERIOD 1000000
on stdcore[2] : out port p_out=XS1_PORT_8D;
void pulse(chanend);
void leddrv(out port, chanend);
int main() {
chan ch;
par{
on stdcore[2] : pulse(ch);
on stdcore[2] : leddrv(p_out,ch);
}
return 0;
}
void leddrv(out port p_lrm, chanend ch0) {
unsigned led=0x01<<3;
unsigned signal;
p_lrm<:led;
while(1) {
ch0:>signal;
p_lrm<:led;
led=(led>>7&0x1)|(led<<1&0xFF);
}
return ;
}
void pulse(chanend ch) {
timer t_mot;
unsigned time;
while(1) {
t_mot:>time;
t_mot when timerafter(time+PWM_PERIOD):>time;
ch<:1;
}
}
```

3.6 Example

To check the procedure of the conversion function, the conversion shown in Fig. 9 was performed as an example. This program outputs a signal to the port of XK–1 and flashes eight LEDs in sequence at a specified time interval (Fig. 10).

Figure 9 shows the actual execution of this program on XK-1. It can be confirmed that the LEDs shown in the upper part of the figure are flashed sequentially. Figure 11 shows an example of conversion from the source file to Hichart. Figure 12 shows the result of conversion from Hichart to XC.

4   Conclusion

In this paper, we proposed a visual embedded software development environment for the XC language for an XMOS evaluation board. We confirmed that bidirectional conversion between Hichart and XC is possible in our proposed system. Therefore, our visual development environment can be used for embedded software development.

References

1. Programming XC on XMOS Devices.: http://www.xmos.com/download/private/Programming-XC-on-XMOS-Devices(1).pdf
2. XMOS. http://www.xmos.com/
3. Yaku, T., Futatsugi, K.: Tree structured flow-chart. In: Memoir of IEICE, pp. AL–78 (1978)
4. Nishino, T.: Attribute graph grammars with applications to hichart program chart editors. Adv. Softw. Sci. Technol. **1**, 89–104 (1989)
5. Vigna, P.D., Ghezzi, C.: Context-free graph grammars. Inf. Control **37**, 207–233 (1978)
6. Adachi, Y., Anzai, K., Tsuchida, K., Yaku, T.: Hierarchical program diagram editor based on attribute graph grammar. In: Proceedings of COMPSAC, vol. 20, pp. 205–213 (1996)
7. Sugita, K., Adachi, A., Miyadera, Y., Tsuchida, K., Yaku, T.: A visual programming environment based on graph grammars and tidy graph drawing. In: Proceedings of The 20th International Conference on Software Engineering (ICSE '98), vol. 2, pp. 74–79 (1998)
8. Miyazaki, M., Ruise, K., Tsuchida, K., Yaku, T.: An NCE attribute graph grammar for program diagrams with respect to drawing problems. IEICE Tech. Rep. **100**(52), 1–8 (2000)
9. Rozenberg, G.: Handbook of graph grammar and computing by graph transformation volume 1. World Scientific Publishing (1997)

Development of an Interface for Volumetric Measurement on a Ground-Glass Opacity Nodule

Weiwei Du, Dandan Yuan, Xiaojie Duan, Jianming Wang,
Yanhe Ma and Hong Zhang

Abstract Although radiologists easily recognize lung nodules in CT volume data, and then judge their benign or malignant based on the type of lung nodules, some lung nodules also are difficult to be detected because of their size or shape and so on such as ground-glass opacity nodules (GGO). Some features of GGO nodules are necessary because they can help radiologists to recognize benign or malignant of GGO nodules such as to find the boundaries in order to obtain the volume of GGO nodules. However, different radiologists can give different boundaries of GGO nodules depended on radiologists' personal habits. It was difficult to obtain the boundaries of GGO nodules which were satisfied with all radiologists. This study is to develop an interface to obtain the boundaries of GGO nodules by using expectation–maximization (EM) algorithm (US Cancer Statistics Working Group. United States cancer statistics: 19992012. Incidence and mortality Web-based report. Atlanta, GA: US Department of Health and Human Services, CDC, National Cancer Institute, 2015, [1]) and the histogram method as radiologists' personal habits because the parameters of the EM algorithm and the threshold values of the histogram method can be adjusted. Experimental results showed the proposed interface can obtain the boundaries of GGO nodules as radiologists' personal habits. This study can reduce the burden of radiologists effectively.

W. Du (✉)
Department of Information and Human Science, Kyoto Institute of Technology,
Kyoto, Japan
e-mail: duweiwei@kit.ac.jp

D. Yuan · X. Duan · J. Wang
School of Electronics and Information Engineering, Tianjin Polytechnic University,
Tianjin, China

Y. Ma · H. Zhang
Tianjin Chest Hospital, Tianjin, China

© Springer International Publishing AG 2018
R. Lee (ed.), *Computer and Information Science*, Studies in Computational
Intelligence 719, DOI 10.1007/978-3-319-60170-0_16

1 Introduction

Lung cancer is still the highest mortality rate of all cancers in many countries. The mortality rate of lung cancer is about 70% in all cancers based on International Agency for Research on Cancer (IARC) [2, 3]. In USA, the American Cancer Society estimates that 224,390 people die of lung cancer in 2016 [4]. In Japan, 77,300 people die of lung cancer [5]. In China, the mortality rate of lung cancer is growing rapidly year by year due to some reasons such as haze and environmental pollution. The mortality of lung cancer is 610,200 cases in 2015 [6]. Lung cancer has a low survival rate, but if tumor of lung has been detected early, the survival rate will be up to 90% in 5 years [7].

Radiologists easily recognize lung nodules in CT volume data, and then judge their benign or malignant based on the type of lung nodules [8]. However, some lung nodules also are difficult to be detected because of their size or shape and so on such as ground-glass nodules. Therefore, this kind of lung nodules need to computer-assisted techniques and pathologic findings to analyze their benign or malignant [9, 10].

This study focuses on ground-glass opacity nodules (GGO) [9]. GGO nodules include part-solid GGO nodules and pure GGO nodules. Part-solid GGO nodules have two parts which are solid part and no solid part. The features of GGO nodules maybe help radiologists to recognize benign or malignant of lung nodules.

In this paper, radiologists' personal habits represent that radiologists give different boundaries of GGO nodules as their personal habits. Thus, it does not exist a standard boundary to a GGO nodule. However, [11] only gives the 3D viewer of a lung nodule as shown in Fig. 1 with the whole volume of the GGO nodules without the volume of solid part. The aim of this study was to develop an interface to obtain the boundaries of GGO nodules by using expectation–maximization (EM) algorithm [1] and the histogram method as radiologists' personal habits. In other words, the boundaries of GGO nodules can be adjusted by using parameters of EM algorithm or the threshold values of histogram method as radiologists' personal habits. Moreover, the proposed interface can obtain the boundaries of the GGO nodule including solid part automatically. Experimental results showed the proposed interface can obtain the boundaries

Fig. 1 3D viewer of a lung nodule by software [11]

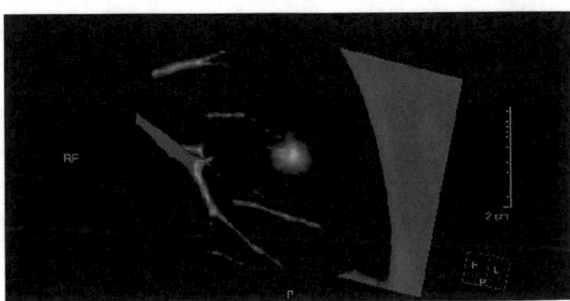

of GGO nodules as radiologists' personal habits. This study can reduce the burden of radiologists effectively.

The remainder of this paper is organized as follows: Section 2 introduces related work on the proposed interface. Section 3 expresses the interface on how to segment lung nodules. Section 4 shows the merits of the interface by the experimental results. Section 5 draws conclusions and further work.

2 Related Work on the Proposed Interface

The proposed interface is designed and provided by [1] using the EM method and the histogram method. The two methods are suitable with Fig. 2. The boundaries of the whole GGO nodule and solid part are obtained by next steps.

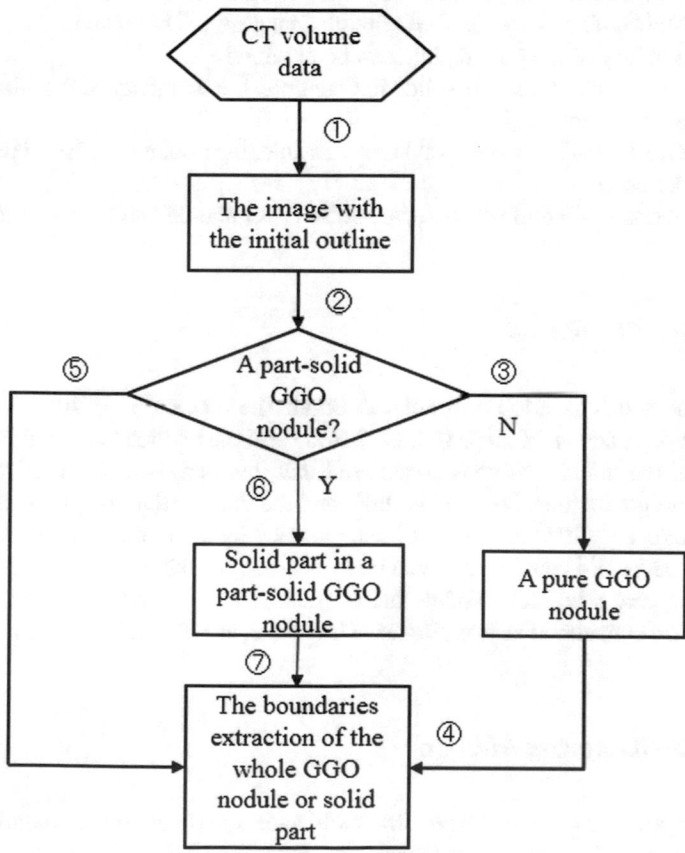

Fig. 2 Flowchart of the boundaries extraction of GGO nodules

(a). The image with the initial   (b). The whole GGO   (c). Solid part in the GGO
        outline   nodule   nodule

Fig. 3 Whole GGO nodule and solid part are segmented by using the EM method

① A slice is selected from the CT volume data.
② A radiologist gives an initial outline including a GGO nodule.
③ The GGO nodule will be judged by a radiologist. If the GGO nodule is not a part-solid GGO nodule, the GGO nodule is a pure GGO nodule.
④ The boundary of the pure nodule can be obtained.
⑤ If the GGO nodule is a part-solid GGO nodule, the boundary of the whole GGO nodule can be obtained.
⑥ If the GGO nodule is a part-solid GGO nodule, the boundary of solid part needs to be obtained.
⑦ The boundary of solid part in a part-solid GGO nodule can be obtained.

2.1 The EM Method

The EM method is used twice in order to obtain the boundaries of the whole GGO nodule which is shown in steps ④ and ⑤ and solid part which is shown in step ⑦. The part of the initial outline is considered that there are two classes. One is that the part does not include the GGO nodule, and the other is that the part includes the GGO nodule. The GGO nodule can be segmented by using the EM method. Solid part in the part-solid GGO nodule can be segmented by using the EM method again because the part-solid GGO nodule includes no solid part and solid part. The two parts also are considered as two classes. The results are shown in Fig. 3.

2.2 The Histogram Method

The histogram method is set by two threshold values to obtain the boundaries of the whole GGO nodule which is also shown in steps ④ and ⑤ and solid part which is

(a). The histogram with the initial outline

(b). The whole GGO nodule

(c). Solid part in the GGO nodule

Fig. 4 Whole GGO nodule and solid part are segmented by using the histogram method

shown in step ⑦. The two threshold values are obtained by Fig. 4a. The results of the whole GGO nodule and solid part are shown in Fig. 4b, c.

3 Development of the Interface on Lung Nodules Segmentation

The proposed interface of this study was developed with MATLAB software on the computer with Intel(R) Core(TM) i3-3110 M CPU2.40 GHZ and RAM: 4.00 GM. The proposed interface is shown in Fig. 5. This interface will be introduced by A, B, C, and D based on different functions. This study introduces two methods to segment the whole GGO nodule and solid part in C and D. One is the EM method, and the other is the histogram method.

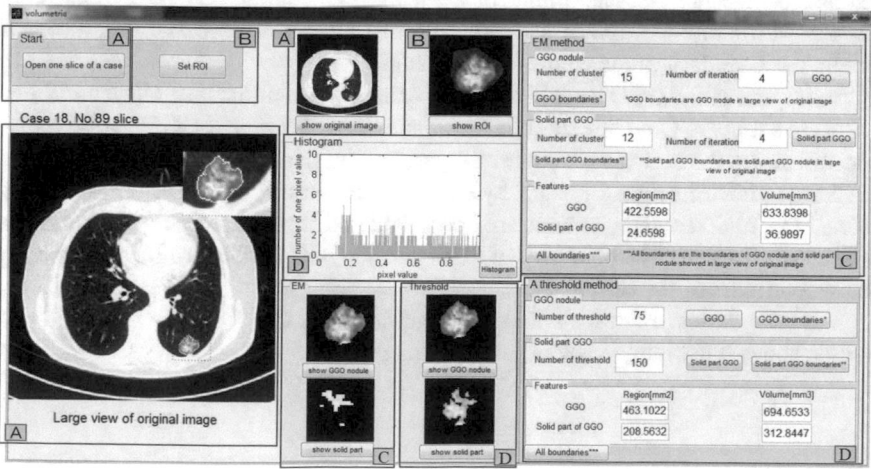

Fig. 5 Proposed interface

3.1   A: Showing One Slice from a CT Volume Data

Radiologists can give an initial outline in "Large view of original image" at random when radiologists click "Set ROI" in the Fig. 5. ROI represents region of interest. Only region of ROI is segmented when radiologists click "show ROI."

3.2   B: Giving the Initial Outline to the Original Image

Radiologists can give an initial outline in "Large view of original image" at random when radiologists click "Set ROI" in the Fig. 5. ROI represents region of interest. Only region of ROI is segmented when radiologists click "show ROI."

3.3   C: Segmenting the Whole GGO Nodule and Solid Part by Using the EM Method

The EM method needs to set some parameters for obtaining the boundaries of the GGO nodules such as the number of cluster and iteration.

The whole GGO nodule can be obtained when number of cluster and iteration are set in GGO nodule. By clicking the button "GGO," the whole GGO nodule is showed above the button of "show GGO nodule." The whole GGO nodule can be enlarged to be observed by clicking the button "show GGO nodule." The whole GGO nodule can be showed in "Large view of original image" by clicking the button "GGO boundaries." At the same time, the region of one slice is computed and showed in features of EM method. The volume of the whole GGO nodule can be computed and showed in the row of GGO of features. The EM method is used as the first time to obtain the boundary of the whole GGO nodule.

Solid part of the GGO nodule can be obtained when number of cluster and iteration are set in GGO nodule. By clicking the button "Solid part GGO," solid part is showed above the button of "show solid part." Solid part of GGO nodule is showed in "Large view of original image" by clicking the button "Solid part GGO boundaries." The region and the volume are computed and showed at the row of solid part of GGO in features. The boundary of solid part in GGO nodule is obtained by using the second time of the EM method.

"All boundaries" represents that the boundaries of GGO nodule and solid part nodule showed in "Large view of original image."

3.4 D: Segmenting the Whole GGO Nodule and Solid Part by Using the Histogram Method

The histogram method needs to set some threshold values parameters for obtaining the boundaries.

The histogram of ROI can be obtained by clicking the button "Histogram." Threshold values of GGO nodules and solid part GGO can be obtained by the wave of histogram. The remainder buttons have the same function with the EM method.

4 Experiments

The institutional review board of Tianjin Chest Hospital, China, approved this study and informed consent was obtained from each patient. The age range of patients is from 26 to 78 years old. The scanning range of CT volume data was from the thoracic inlet to the lung bottom. Twenty-one GGO nodules were applied with the proposed interface of this study.

Some parameters of CT volume data were shown as follows: The resolution of one slice is 512×512 pixels. The thickness of one slice is 1.5 mm. Only the thickness of one slice to one patient is 5.0 mm because the CT volume data will be given the patient by film form. The pixel size is 0.59–0.93 mm. Tube voltage is 120–130 KV, and tube current is 39–450 mA. Tube voltage and tube current are adjusted depended on the weight of patients. The range of field of view (FOV) is 300–446 mm. The FOV is adjusted depended on chest circumferences of patients.

4.1 The Boundaries of GGO Nodules Extraction Automatically

The boundaries of the whole GGO nodule and solid part can be obtained by using EM method with different parameters such as the number of clusters and iteration or by using threshold values based on the histogram method. The proposed interface took a few seconds to obtain the boundaries depended on radiologists' personal habits to one slice. Radiologists spent 20 minutes to obtain the boundary of the whole GGO nodule to one slice by hand which is shown in Fig. 6.

4.2 Following Radiologists' Willing to Give Boundaries

As we all know, different radiologists have their different habits to give the boundaries of the whole GGO nodule and solid part. Although the boundaries are different, they are similar. Therefore, the proposed interface of this paper can obtain the bound-

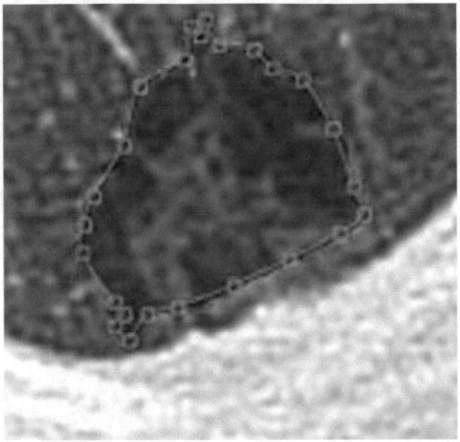

Fig. 6 Boundary of the whole GGO nodule by hand

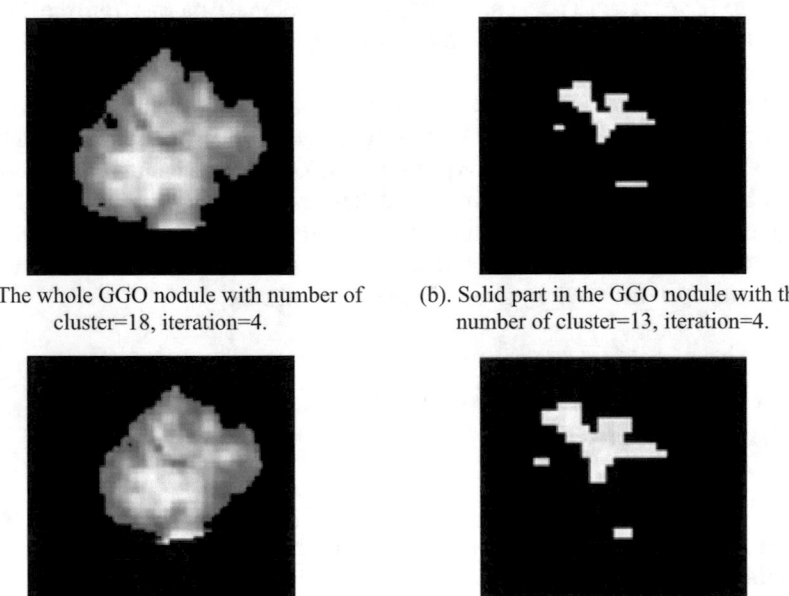

(a). The whole GGO nodule with number of   (b). Solid part in the GGO nodule with the
cluster=18, iteration=4.   number of cluster=13, iteration=4.

(c). The whole GGO nodule with number of   (d). Solid part in the GGO nodule with number
cluster=15, iteration=4.   of cluster=12, iteration=4.

Fig. 7 EM method with different number of cluster and iteration

aries of the whole GGO nodule and solid part what radiologists wanted by setting
different parameters. The acquisition of these boundaries takes only a few seconds.
Figure 7 shows the results of the EM method with different number of cluster and
iteration. Figure 8 shows the results of the histogram method with different threshold
values.

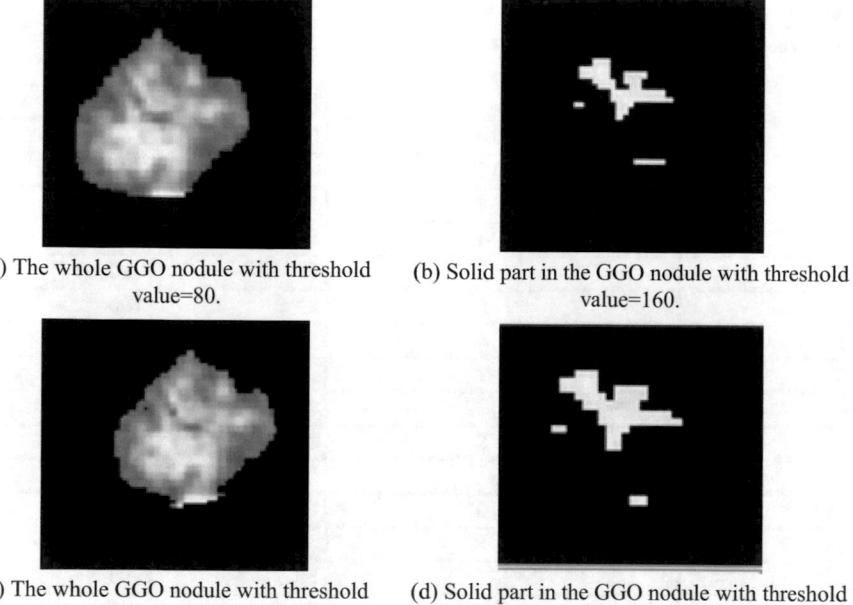

(a) The whole GGO nodule with threshold value=80.

(b) Solid part in the GGO nodule with threshold value=160.

(c) The whole GGO nodule with threshold value=75.

(d) Solid part in the GGO nodule with threshold value=150.

Fig. 8 Histogram method with different threshold values

4.3 Evaluation on the Proposed Interface

Twenty-one GGO nodules were experimented with the proposed interface. Table 1 represents successful segmentation rates with 21 cases. In Table 1, "s" represents part-solid GGO nodules and "p" represents pure GGO nodules in type of the GGO nodules. The number of slices represents the number of the slices including GGO nodules in a CT volume data, in other word, a case. Formula (1) represents a calculation method on a successful segmentation rate. In formula (1), number of segmentation slices represents the number of the slices including the GGO nodules which are segmented by using the EM method or the histogram method. Therefore, "Successful segmentation rate" represents the following formula:

$$Successful \quad segmentation \quad rate(\%) = \frac{number \quad of \quad segmentation \quad slices}{number \quad of \quad slices} \times 100\% \tag{1}$$

From the results of "Successful segmentation rate" in Table 1, the histogram method is better than the EM algorithm. However, the results only explained that the parameters of the histogram were easier to be adjusted than the EM method's parameters. "Successful segmentation rate" cannot be used as a basis to explain which method

Table 1 Successful segmentation rates with 21 cases

No. in cases	Type of the GGO nodules	Number of slices	Successful segmentation rate %	
			EM (%)	Histogram (%)
1	s	16	81	100
2	s	10	80	90
3	s	4	75	100
4	s	14	93	93
5	s	5	80	80
6	s	6	83	100
7	p	4		
8	s	7	86	86
9	s	13	85	92
10	s	8	88	88
11	s	11	91	100
12	s	3	67	100
13	p	8		
14	s	17	88	94
15	s	18	89	94
16	s	3	67	100
17	s	11	91	91
18	s	15	87	93
19	s	5	80	100
20	p	8		
21	s	17	88	94

is better. "Successful segmentation rate" in case 7, 13, and 20 are blanks, because the three cases are pure GGO nodules. The parameters adjustment of the EM method and the histogram method was executed by a master student. But experimental results from Table 1 showed the proposed interface can obtain the boundaries of GGO nodules based on adjustment of parameters. Therefore, the proposed interface is helpful for radiologists to obtain the boundaries what they wanted. Thus, this study can reduce the burden of radiologists effectively.

5 Conclusions

This paper proposed an interface to obtain the boundaries of the whole nodule and solid part as radiologists' personal habits. The boundaries can be obtained automatically and rapidly. Therefore, the proposed interface can reduce some burden of radiologists and is helpful to judge benign or malignant of lung nodules.

The proposed interface should be improved as the following problems in the further.

(a) The scope of the parameters of EM method should be determined in order to let radiologists are easy to select the parameters in the proposed interface.
(b) Which slice exists GGO nodule should be recognized automatically in order to compute the volume of GGO nodules easily.

References

1. Miao, Y., Wang, J., Du, W., Ma, Y., Zhang, H.: Volumetric measurement of ground-glass opacity nodules using expectation-maximization algorithm. In: The 4th IIAE International Conference on Intelligent Systems and Image Processing, Kyoto, Japan, pp. 317–321 (2016)
2. US Cancer Statistics Working Group.: United States cancer statistics: 19992012. Incidence and mortality web-based report. Atlanta, GA: US Department of Health and Human Services, CDC, National Cancer Institute (2015)
3. Horn, L., Pao, W., Johnson, D.H., Kasper, D.L., Jameson, J.L., Fauci, A.S., Hauser, S.L., Loscalzo: Harrison's Principles of Internal Medicine, 18th ed. McGraw-Hill (2012). ISBN: 0-07-174889-X
4. Rebecca, L., Siegel, K.D., Jemal, A.: Cancer statistics. CA: A Cancer J. Clinic. **66** (1), 7–30 (2016)
5. Katanoda, K., Kamo, I.K., Saika, K., et al.: Short-term projection of cancer incidence in japan using an age-period interaction model with spline smoothing. Jpn. J. Clinic. Oncol. **44**(1), 36–41 (2014)
6. Chen, W., Zheng, R., Baade, P.D., et al.: Cancer statistics in China. CA: A Cancer J. Clinic. **66**(2), 115–132 (2016)
7. Elsayed, O., Mahar, K., Kholief, M., et al.: Automatic detection of the pulmonary nodules from CT images. In: SAI Intelligent Systems Conference, London, UK, pp. 742–746 (2015)
8. Travis, W., et al.: IASLC/ATS/ERS: international multidisciplinary classification of lung adenocarcinoma. J. Thorac. Oncol. **6**(2) (2011)
9. Chae, H.-D., Park, C.M., Park, S.J., Lee, S.M., Kim, K.G., Goo, J.M.: Computerized texture analysis of persistent part-solid ground-glass nodules: differentiation of preinvasive lesions from invasive pulmonary adenocarcinomas. Radiology **273**(1), 285–293 (2014)
10. Ko, J.P., Suh, J., Ibidapo, O., Escalon, J.G., Li, J., Pass, H., Naidich, D.P., Crawford, B., Tsai, E.B., Koo, C.W., Mikheev, A., Rusinek, H.: Lung adenocarcinoma: correlation of quantitative CT findings with pathologic findings. Radiology **280**(no. 3), 931–939 (2016)
11. DICOM Conformance Statement for DICOM Viewer Release 3.0, Philips Healthcare (2013)

Efficient Similarity Measurement by the Combination of Distance Algorithms to Identify the Duplication Relativity

Manop Phankokkruad

Abstract This paper studied the efficient similarity measurement in order to improve the duplication detection techniques in a programming class. This work used the combination of the three proficient algorithms that include Smith-Waterman, longest common subsequence, and Damalau-Levenshtein distance to measure the distance between each pair of the code files. In order to identify the proximity of the person who duplicates each other, this work applied the frequency of occurrence technique to score the accumulated similarity of duplication, the number of times or the regularity of the duplication happens, and the relationship between incidence and time period. Moreover, this work proposed the technique to identify a group of people positioned closely together. The result shows that the proposed of the combination of algorithms could measure the similarity of files efficiently. This work could identify the relativity of a person who frequently duplicates together, and the people positioned closely together. Finally, this work suggested the number of time that student duplicated the code files.

1 Introduction

A rapidly increasing of large data volume from the variety of sources, the duplicates are difficult to detect. Usually, the duplicate detection methods are included in almost every data warehouse. Duplicate detection is the process of perceiving the multiple representations of same objects. This problem has been widely inspected in many research literature [4]. The duplication detections are motivated by the impact duplicates to have in many variety applications. There are several techniques have been proposed to detect the duplication, such as textual comparison [3], structural comparison using pattern matching [14], code fingerprints [10], and statistical analysis of the code [11]. An interesting case that occurs in the duplication detection is applied in text file comparing. The computer programming courses need many

M. Phankokkruad (✉)
Faculty of Information Technology, King Mongkut's Institute
of Technology Ladkrabang, Bangkok 10520, Thailand
e-mail: manop@it.kmitl.ac.th

© Springer International Publishing AG 2018   219
R. Lee (ed.), *Computer and Information Science*, Studies in Computational
Intelligence 719, DOI 10.1007/978-3-319-60170-0_17

assignments to get the good strong problem-solving skills. Since these assignments usually are program codes which are the text, thus it can be copied to the others easily. Many students with a little motivation enroll in programming courses to meet the course requirement. These students do not intend to work hard enough to do the programming exercises. They simply want to get over it by any easy means [15]. Consequently, there are several clone detection tools and technique [1], many prior works have addressed the problem of identifying duplicate code, and the techniques to detect the copying by tools. For instance, Wang et al. [18] proposed the preventing and detecting plagiarism in programming courses. They described a tool to support the instructor find the suspecting codes that aim to reduce the plagiarism and improve the ability of the student programming. The analysis of code to identify duplication and plagiarism is extensive. An interesting case of detecting student code plagiarism using statistical comparisons was suggested by Jankowitz [6]. This work used the static execution tree of a program to determine a fingerprint of the program. Likewise, Johnson [7] used a specific heuristic to gather a number of lines of source code on which applied the fingerprint algorithm. The existing approaches have focused on efficient algorithms for locating potential duplicates rather than precise similarity metrics for comparing records. Furthermore, they present the methods for improving duplicate detection accuracy using any method [12]. In particular, a duplication detection is fully considered as one of the most significant applications [2]. Actually, the duplication detection is the process of measuring and calculating the degree of a similarity between two or more text objects. Many researchers have discussed similarity in contexts of copy detection. Steinberger et al. [17] combined the algorithm with the other method in the analysis of the similarity between texts. Unfortunately, there is no research suggests about how to find the relationship between the texts from the students. Because each algorithm represents the specific function and different purpose.

Therefore, the goal of this paper intends to study the alternative techniques for implementing by the combination of distance algorithms that improve the efficiency of the similarity measurement. This performed techniques are not only appropriate for detecting the similarity between the duplicate files, but also this method can help to identify the relativity of the person in the group who duplicated the code each other. Moreover, this work applied the frequency of occurrence technique [13] to score the accumulated similarity of duplication, the number of times or the regularity of the duplication happens, and the relationship between incidence and time period.

The remainder of this paper is organized as follows: Section 2 provides the problem definition, the duplication detection processes. Section 3 presents theoretical basis, the proposed method and the details of the algorithms. Section 4 describes the experiments. Section 5 describes the results and evaluation of the research. Finally, the summary and conclusion are presented in Sect. 6.

2 Proposed Algorithms and Methods

In this section, this work defines the state of the problem, basic concepts of duplication detection, similarity measurement, and related algorithms. The starting point is the problem of duplication detection, and how to overcome.

2.1 Problem Definition

The computer programming courses need many assignments to get the good strong problem-solving skills. These are one of those things students get much better with practice. Since these assignments usually are program codes, which are plenty the text, thus it can be copied to the others easily. The students who copy the code from the others, they lack to learn programming and lost the chance to practice themselves. Many students with a little motivation enroll in programming courses to meet the course requirement. These students do not intend to work hard enough to do the programming exercises as required for learning and developing programming skill. They simply want to get over it by any easy means [15]. For this reason, the plagiarism in programming assignments frequently occurs. In a big class, there are a lot of exercises required a huge number of comparisons, which makes it is very difficult for the instructor manually inspect all the programming codes. Therefore, an efficient automatic detecting technique for programming code duplication is required.

Since the programming code is the text file that contains a sequence of characters, this code can be compared with the calculation and display of the differences and similarities between code files. In this case, this work applies the Smith-Waterman algorithms to measure the similarity between code files. To extract a part of copied code, the longest common subsequence (LCS) algorithm is adopted. Because sometimes students who duplicated the code from the other may modify a little of the gained code to cheating. Thus, to make sure that code is the exactly duplicated code file, the Damalau-Levenshtein distance is used to count the minimum number of keystoke operations that edit the original file into the new one. From the prior study, there is already many automatic program plagiarism detecting system but there is no tool that identifies the relationship between the group of a person who copies the codes each other. Therefore, this work proposed a technique to identify this relationship by the frequency of occurrence technique into the Euclidean distance matrix.

2.2 Duplication Detection Processes

This work designs the processes of duplication detection by means of the problem and solving. Figure 1 shows a proposed of the duplication detection processes. It consists of five steps: file database collection, calculated the similarity, apply the similarity thresholds, calculate the frequency of occurrence, and create the distance matrix. The detail of each step is described as follows:

Fig. 1 A duplicate
detection process diagram

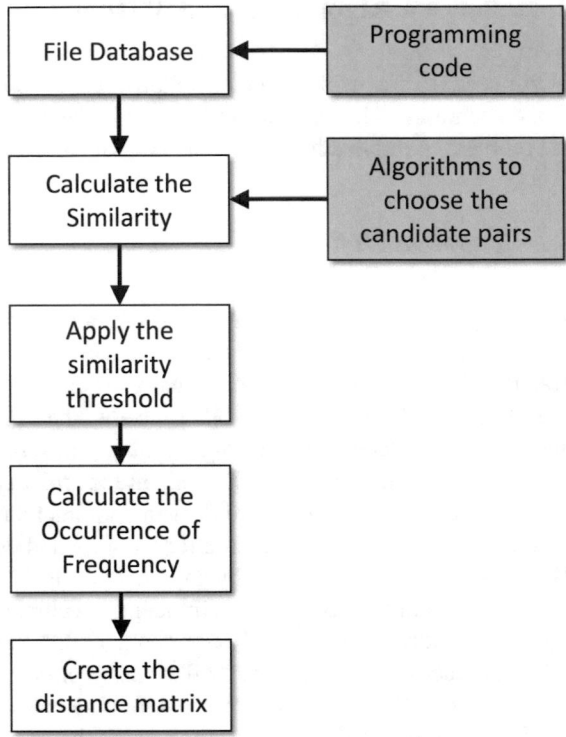

Firstly, this work collects the set of files, that is, suspected to contain duplicates come from the programming class. Then, an algorithm chooses to promise candidate pairs that have a sufficiently high similarity at first glance by the first algorithm. These candidate pairs are the input to measure the more calculated similarity and edit distance. Then, all of results are applied the similarity thresholds to eliminate the minor value for the similarity pair. Furthermore, it is calculated with frequency of occurrence technique that meets the high-similarity value be between the pairs. This step conducts by giving the score that related to the similarity. Finally, all of the similarity create the distance matrix to decide whether the pair is indeed a duplicate or not. Moreover, this step will identify the relationship between a person and a person, a person and a group, who often copies the code each other.

3   Combination of Distance Algorithms

This section presents the theoretical basis of the algorithms that it is used as the similarity and distance measurement. Since, this work combines the several distance algorithms, each algorithm is used in the different purpose. The details of each algorithm are described to follow.

3.1 Smith-Waterman Algorithm

Smith-Waterman algorithm [16] is a well-known algorithm for performing local sequence alignment. It is used for determining similar regions between two nucleotides, that is, for determining similar regions between two strings. Moreover, the Smith-Waterman algorithm compares segments of all possible lengths and optimizes the similarity measure between two strings. The Smith-Waterman algorithm consists of two steps include the calculation of the similarity matrix score and tracing back of the similarity matrix to examine the optimal alignment. In the first step, it will spend the largest part of the total calculation time. For two sequences S and T, where the length of S is n, $S = n$. Let the length of T is m, $T = m$. Let $V(i,j)$ denote the optimal alignment score of two subsequence $S[1]...S[i]$ and $T[1]...T[j]$, and the computation of $V(i,j)$ is defined as Eq. (1).

$$
V(i,j) = max \begin{cases} 0 \\ V(i-1,j-1) + \sigma(S[i], T[j]) \\ V(i-1,j) + \sigma(S[i], -) \\ V(i,j-1) + \sigma(-, T[j]) \end{cases}
\tag{1}
$$

where k = deletion of length k, l = deletion of length l, and Wk and Wl is the gap cost function.

In Eq. (1), let "$-$" denote a null character; $V(i,0)$ is the result of comparing each character in S with a null character in T; where $V(0,j)$ is the counterpart of comparing each character in T with a null character in S; and $(S[i], T[j])$ denote the value of a substitution matrix. As computing the similarity matrix, ever the score of any matrix element $V(i,j)$ depends on the score of three other elements; the up-left neighbor element $V(i-1,j-1)$, the left neighbor $V(i,j-1)$, and the up to neighbor $V(i-1,j)$.

3.2 Longest Common Subsequence

The longest common subsequence (LCS) [9] is an algorithm on the basis of data comparison. LCS has wide applications in the bioinformatics research and also widely used by revision control systems. Generally, LCS is the algorithm of finding the longest subsequence common to all sequences in two sequences. Let two test sequences be defined as follows: $X = (x_1, x_2, ..., x_m)$ and $Y = (y_1, y_2, ..., y_n)$. The prefixes of X are $1, 2, ..., m$; the prefixes of Y are $1, 2, ..., n$. Let $LCS(X_i, Y_j)$ represent the set of longest common subsequence of prefixes X_i and Y_j. This set of sequences is given by the following.

$$c[i,j] = \begin{cases} 0 & \text{if } i = 0 \text{ or } j = 0 \\ c[i-1,j-1]+1 & \text{if } x_i = y_j \\ max(c[i,j-1], c[i-1,j]) & \text{if } x_i \neq y_j \end{cases} \quad (2)$$

Here, we have used the tabular notion $c[i,j]$ to denote $r(X[1...i], Y[1...j])$. From Eq. (2), it follows that only the entries $c[i,j]$ such that $[i,j] \in M$ are useful. Therefore, we can ignore all $c[i,j]$ with $[i,j] / \in M$ from the calculation [5]. In the paper, the role of LCS is used for finding the longest subsequence common to all code in a large set of programming code. Then, this algorithm can be subtracted the part of code, which is duplicated.

3.3 Damalau-Levenshtein Distance

Damerau-Levenshtein [8] distance is a string metric for determining the edit distance between two sequences. The distance determines by counting the minimum number of the operations required to transform one string into the other. In addition, it allows modifying this distance by including transpositions of adjacent symbols produce a different distance measure. To indicate the Damerau-Levenshtein distance between two strings, $d_{a,b}(i,j)$ function is defined, whose value is a distance between an i-symbol prefix (or initial substring) of string a and a j-symbol prefix of b. The function is defined recursively as follows:

$$d_{a,b}(i,j) = \begin{cases} max(i,j) \\ min \begin{cases} d_{a,b}(i-1,j)+1 \\ d_{a,b}(i,j-1)+1 \\ d_{a,b}(i-1,j-1)+1 \\ d_{a,b}(i-2,j-2)+1 \end{cases} \\ min \begin{cases} d_{a,b}(i-1,j)+1 \\ d_{a,b}(i,j-1)+1 \\ d_{a,b}(i-1,j-1)+1 \end{cases} \end{cases} \quad (3)$$

where $1_{(a_i \neq b_j)}$ is the indicator function equal to 0 (zero) when $a_i = b_j$ equal to 1 otherwise. Each recursive call matches one of the cases covered by the Damerau-Levenshtein distance: where $d_{a,b}(i-1,j)+1$ corresponds to a deletion from a to b, $d_{a,b}(i,j-1)+1$ corresponds to an insertion from a to b, $d_{a,b}(i-1,j-1)+1_{(a_i \neq b_j)}$ corresponds to a match or mismatch, depending on whether the respective symbols are the same, and $d_{a,b}(i-2,j-2)+1$ corresponds to a transposition between two successive symbols.

3.4 Euclidean Distance Matrix

Euclidean distance is the well-known distance measurement, which transforms the relative of distance into the matrix. This work applies the accumulate of the computational model, which shows how many scores that students duplicate frequently and the student positioned closely together. There is a better way to understand the relationship of these students. Thus, this work creates the accumulated similarity, which measures by the three major algorithms. The relation of the similarity will be transformed into the distance matrix by the Euclidean distance function that even provides the total distance, also represents the score of relationship between students.

To make sure that the code file is duplicated from another student. This work defines the accumulate the total distance by considering the importance of three factors: normal similarity, the ratio of subsequence percentage, and the number of the operations required to transform one string into the other. Firstly, the normal similarity between files, which is calculated by Smith-Waterman algorithm. The ratio of subsequence percentage is the highest percentage of the substring in the file. If it is large, a part of the string in this file may be duplicated. In the case of the number of the operations, this value is evaluated by the Damerau-Levenshtein distance. If the number of operation is small, it means that the file is exactly duplicated. Finally, this work accumulates the total distance by the Euclidean distance function as shown in Eq. (6). The equation shows the distance between file i and j. To denote the distance between vectors i and j which can use the notation $d_{i,j}$. So, this last result can be written as in Eq. (6).

$$L_i = \frac{c\left[i,j\right]}{Total\ size} \times 100 \tag{4}$$

where $c\left[i,j\right]$ denote a distance of longest subsequence which produced by Eq. (2).

$$D_i = \left(1 - \frac{d_{a,b}\left(i,j\right)}{Total\ size}\right) \times 100 \tag{5}$$

where $d_{a,b}\left(i,j\right)$ denote a distance between two strings which produced by Eq. (3).

$$d_{i,j} = \frac{\left(S_i + L_i + D_i\right)}{3} \tag{6}$$

where S_i are the distance that measure by Eq. (1). L_i are the distance of the ratio of subsequence percentage which calculate by Eq. (eq:LCS) and Eq. (5). D_i are the distance of the number of the operations in percentage unit which produced by Eqs. (3) and (5).

$$d_{i,j} = \begin{cases} 0 & \text{if } i = j \\ 0 & \text{if } i > j \\ d_{i,j} & \text{if } i \neq j \text{ and } i < j \end{cases} \tag{7}$$

In order to decrease the amount of calculations and reduce the complexity of the matrix. This considered any function $d_{i,j}$ which satisfies the following conditions (metric axioms) for all points. This work defines the metric axioms in three conditions as following. If the two points differ, that is, $i > j$, then $d_{i,j} > 0$. According to the symmetry axiom $d_{i,j}$ is zero, where $i > j$, $d_{i,j} = d_{j,i}$, that is, the direction of measurement is immaterial. Also, in the case of two points coincide $d_{i,j}$ is zero, that is, $i = j$. For instance, the relation of $d_{1,2}$ is equal to $d_{2,1}$ and also $d_{1,1}, d_{2,2}, \dots, d_{i,i}$ refer to the relation to itself. As described, this work can generate the diagonal matrix of the relation which is illustrated in Eq. (8).

$$E = \begin{bmatrix} 0 & d_{1,2} & d_{1,3} & d_{1,4} & \dots & d_{1,j} \\ 0 & 0 & d_{2,3} & d_{2,4} & \dots & d_{2,j} \\ 0 & 0 & 0 & d_{3,4} & \dots & d_{3,j} \\ 0 & 0 & 0 & 0 & \dots & d_{5,j} \\ \vdots & \vdots & \vdots & \vdots & \ddots & \vdots \\ 0 & 0 & 0 & 0 & \dots & 0 \end{bmatrix} \tag{8}$$

where E is an Euclidean distance matrix.

4 Experiments

This work conducted the experiments by collecting the programming code files from the object-oriented programming (OOP) class. This class consists of 135 students, 10 weeks of the laboratory, and each week comprise of five exercises. Thus, this work will be measured the 134 pair of code files on each exercise, 670 pairs per a week, and 6,700 pairs in the experiment, respectively. This work defines the difference similarity threshold between 70 and 75%, that it depends on the character of each exercise.

The measurement process of this experiment consists of five steps. Firstly, every student code file will be evaluated the similarity distance by the Smith-Waterman algorithm as described in Eq. (1). Furthermore, its will be measured the distance of the longest size of the substring in the code by the LCS algorithm, that it is described in Eq. (2). Likewise, the student files will be evaluated the edit distance between files by the Damalau-Levenshtein distance which is described in Eq. (3). In the second step, the calculated similarity will be measured the accumulated similarity, and also built the Euclidean matrix of each week. Moreover, the similarity value that higher than the threshold will be counted the number of the frequency of occurrence. This number means the numbers of time that the student copied the code from someone else frequently. Likewise, the accumulated similarity of a class will be measured and also built the Euclidean matrix. A number of the duplication will be counted the frequency of occurrence as well.

Finally, all similarity results would be created the cumulative distance matrix by the Euclidean distance in order to measure the score of the relationship by the frequency of occurrence technique. As shown in Eq. (8), this work creates the distance matrix of each exercise, In the same way, the cumulation of a week exercise and course exercise is generated to the cumulation of the distance matrix.

5 Results and Evaluations

This section presents the results obtained when the duplicate detections were applied to all pairs of codes. Furthermore, the methods to evaluate the correctness of the algorithms. The difference similarity measurement can be evaluated in order to prove the efficiency.

From the experiment, the similarity scores and distance between each pair of files were computed using Smith-Waterman, LCS, and Damalau-Levenshtein distance, which were described in Eqs. (1), (2), and (3), respectively. A part of the results are shown in Table 1.

The LCS algorithm gave a part of text which identical to the original source, which is depicted in Fig. 2. Likewise, the similarity score is defined in by the LCS algorithm that was described in Eq. (2). The results of similarity measurement of an exercise by each algorithm are depicted in Table 1. Then, this value is calculated by Eq. (6) to report these distance in the Euclidean matrix. This work does repeat this calculation for all of samples by Eqs. (6) and (7). The distance can be calculated and reported in the Euclidean matrix of an exercise, a week, and ten weeks, respectively. The Euclidean matrix of the accumulated distance of 49 students in a week is shown in Table 2 and Fig. 3. Every time the score is equal or greater than threshold

Table 1 A part of similarity score that is measured by the different algorithms

ID1	ID2	LCS[a]	SW[b]	DLD[c]	ED[d]
019	046	53.77	34.34	333	339.06
019	067	68.32	43.03	147	167.71
019	074	72.34	48.23	151	174.24
019	085	81.32	62.17	104	145.93
019	106	76.36	67.85	140	173.30
019	115	76.26	47.03	161	184.25
019	116	77.07	63.83	134	167.24
…	…	…	…	…	…
019	011	77.54	48.94	119	150.23

[a]Longest Common Subsequence
[b]Smith-Waterman
[c]Damalau-Levenshtein Distance
[d]Euclidean Distance

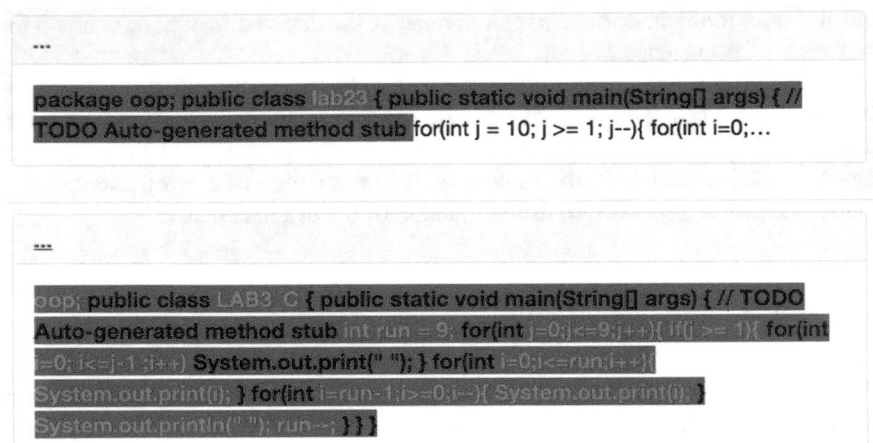

Fig. 2 An example of a part duplicate code which extracts by the LCS algorithm

Table 2 An example of Euclidean distance matrix of 49 students

ID	1	2	3	4	5	6	...	49
1	0	211.9	234.1	246.1	257.0	298.7	...	155.3
2	0	0	241.7	250.6	190.3	252.9	...	108.5
3	0	0	0	349.9	226.7	278.2	...	146.9
4	0	0	0	0	221.7	284.2	...	159.6
5	0	0	0	0	0	236.5	...	135.0
...	0	0	0	0	0	...	...	174.8
...	0	0	0	0	0	...	...	142.7
49	0	0	0	0	0	0	0	0

of each exercise, it is counted the frequency that the student duplicates the code file. Then, the accumulated frequency is created the score matrix as shown in Table 3. An example of the frequency of occurrence score of 20 exercises is shown in Fig. 4. Exactly, both accumulated score of the frequency of occurrence and the accumulated distance reveal the relativity between students who duplicate each other. The closely accumulated score represents the student who is in the same group, and who often copies the code. An example of an accumulated score is presented in Table 4.

In order to evaluate the accuracy of the detection algorithms, this work created the validation set by defining ten pairs of known similarity files, and it was cloned to three sets. These three set files were measured the distance by the three kinds of the distance algorithms. Then, it was compared with the calculated results from the reliable Unix tools: *sdiff*, *diff*, and *diffstat*. This tool provides the useful library that can be integrated into the application and command line. The Smith-Waterman algorithm was evaluated by *sdiff* utility, and the LCS algorithm was evaluated by

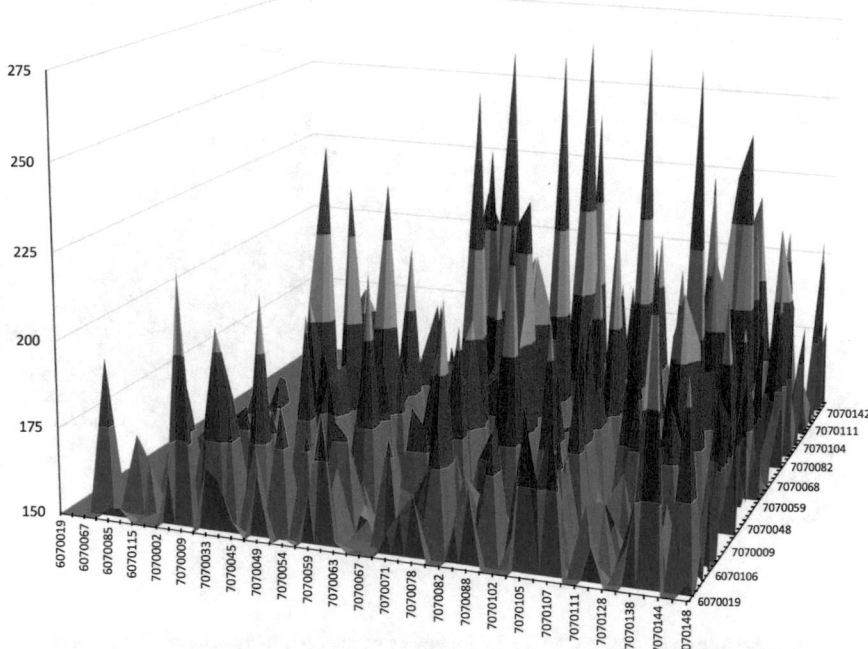

Fig. 3 An example chart of duplication score in the distance matrix by a week

Table 3 An example of the score matrix by the frequency of occurrence technique of 49 students in 50 exercises

ID	1	2	3	4	5	6	...	49
1	0	42	44	46	47	47	...	32
2	0	0	37	33	26	24	...	23
3	0	0	0	48	18	18	...	14
4	0	0	0	0	10	21	...	11
5	0	0	0	0	0	22	...	16
...	0	0	0	0	0	...	...	25
...	0	0	0	0	0	...	...	13
49	0	0	0	0	0	0	0	0

diff utility. In that way, the Damalau-Levenshtein distance was evaluated by *diff* and *diffstat* utility. Since, the results from the three Unix tools are measured separately, it was combined and calculated by Eq. (6). The evaluation shows that the results from three algorithms are same as the results from the three Unix tools on both validation set and actual test set.

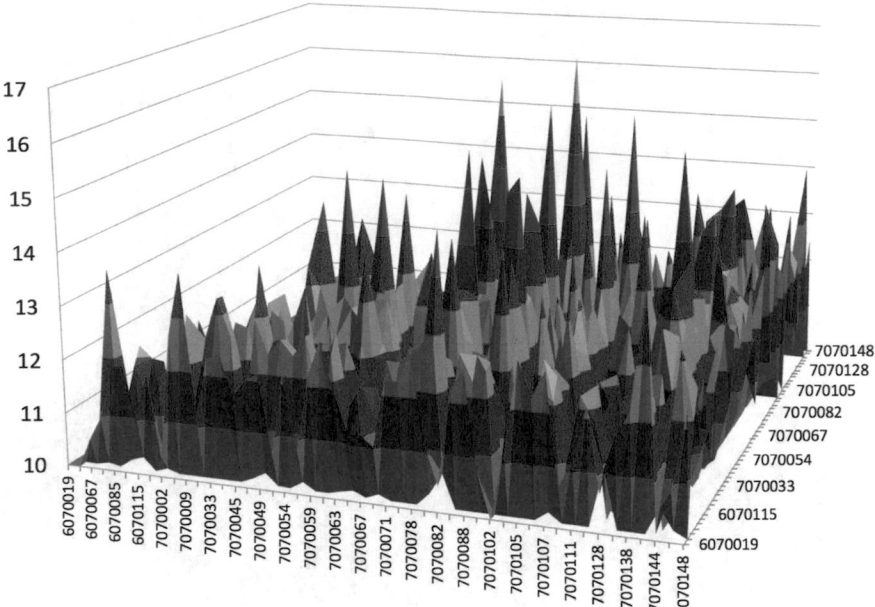

Fig. 4 An example chart of score by the frequency of occurrence technique of 20 exercises

Table 4 An example of the accumulate distance which shows the relativity between person

ID1	ID2	Accu. distance
..019	..046	349.9
..019	..067	298.7
..019	..074	284.2
..019	..085	278.0
..019	..106	267.2
..019	..115	259.9
..019	..116	258.8
..019	..009	258.3
...	...	...
..019	..011	251.1

6 Conclusion

This paper studied the efficient similarity measurement in order to improve the duplication detection techniques in programming class. This work used the combination
of the three proficient distance algorithms that include Smith-Waterman, longest
common subsequence, and Damalau-Levenshtein distance to measure the distance
between each pair of the code files. In order to identify the proximity of the person

who duplicates each other, this work applied the frequency of occurrence to score the accumulate the number of time in duplication, the relationship between incidence, and time period. Moreover, this work proposed the technique to identify the people who positioned closely together. The result reveals that the proposed of the combination of algorithms could measure the similarity of files efficiently. This work could identify the relativity of a person who frequently duplicates together, and the people positioned closely together. Finally, this work suggested the number of time that student duplicated the code files.

References

1. Bellon, S., Koschke, R., Antoniol, G., Krinke, J., Merlo, E.: Comparison and evaluation of clone detection tools. IEEE Trans. Softw. Eng. **33**(9), 577–591 (2007)
2. Brin, S., Davis, J., García-Molina, H.: Copy detection mechanisms for digital documents. In: Proceedings of the 1995 ACM SIGMOD International Conference on Management of Data. SIGMOD '95, pp. 398–409. ACM, New York, NY, USA (1995)
3. Ducasse, S., Rieger, M., Demeyer, S.: A language independent approach for detecting duplicated code. In: IEEE International Conference on Software Maintenance. (ICSM '99) Proceedings, pp. 109–118 (1999)
4. Elmagarmid, A., Ipeirotis, P., Verykios, V.: Duplicate record detection: a survey. IEEE Trans. Knowl. Data Eng. **19**(1), 1–16 (2007)
5. Iliopoulos, C.S., Rahman, M.S.: A new efficient algorithm for computing the longest common subsequence. Theory Comput. Syst. **45**(2), 355–371 (2009)
6. Jankowitz, H.T.: Detecting plagiarism in student pascal programs. Comput. J. **31**(1), 1–8 (1988)
7. Johnson, J.H.: Identifying redundancy in source code using fingerprints. In: Proceedings of the 1993 Conference of the Centre for Advanced Studies on Collaborative Research: Software Engineering—Volume 1, CASCON '93, pp. 171–183. IBM Press (1993). http://dl.acm.org/citation.cfm?id=962289.962305
8. Levenshtein, V.I.: Binary codes capable of correcting deletions, insertions and reversals. Sov. Phys. Dokl. **10**(8), 707–710 (1966)
9. Maier, D.: The complexity of some problems on subsequences and supersequences. J. ACM **25**(2), 322–336 (1978)
10. Manber, U.: Finding similar files in a large file system. In: Proceedings of the USENIX Winter 1994 Technical Conference on USENIX Winter 1994 Technical Conference, WTEC'94, pp. 2–2 USENIX Association, Berkeley, CA, USA (1994)
11. Mayrand, J., Leblanc, C., Merlo, E.M.: Experiment on the automatic detection of function clones in a software system using metrics. In: 1996 Proceedings of International Conference on Software Maintenance, pp. 244–253 (1996)
12. Murakami, H., Hotta, K., Higo, Y., Igaki, H., Kusumoto, S.: Gapped code clone detection with lightweight source code analysis. In: 2013 21st International Conference on Program Comprehension (ICPC), pp. 93–102 (2013)
13. Nardo, F.D., Agostini, V., Knaflitz, M., Mengarelli, A., Maranesi, E., Burattini, L., Fioretti, S.: The occurrence frequency: a suitable parameter for the evaluation of the myoelectric activity during walking. In: 2015 37th Annual International Conference of the IEEE Engineering in Medicine and Biology Society (EMBC), pp. 6070–6073 (2015)
14. Paul, S., Prakash, A.: A framework for source code search using program patterns. IEEE Trans. Softw. Eng. **20**(6), 463–475 (1994)
15. Rosales, F., Garcia, A., Rodriguez, S., Pedraza, J.L., Mendez, R., Nieto, M.M.: Detection of plagiarism in programming assignments. IEEE Trans. Educ. **51**(2), 174–183 (2008)

16. Smith, T., Waterman, M.: Identification of common molecular subsequences. J. Mol. Bio. **147**(1), 195–197 (1981)
17. Steinberger, R., Pouliquen, B., Hagman, J.: Cross-lingual document similarity calculation using the multilingual thesaurus eurovoc. In: Gelbukh, A. (ed.) Computational Linguistics and Intelligent Text Processing. Lecture Notes in Computer Science, vol. 2276, pp. 415–424. Springer, Berlin Heidelberg (2002)
18. Wang, C., Liu, Z., Liu, D.: Preventing and detecting plagiarism in programming course. Int. J. Secur. Appl. 269–278 (2013)

Author Index

© Springer International Publishing AG 2018
R. Lee (ed.), *Computer and Information Science*, Studies in Computational
Intelligence 719, DOI 10.1007/978-3-319-60170-0